The Impact of Global Warming on Texas

The Impact of Global Warming on Texas

Second edition

Edited by Jurgen Schmandt, Gerald R. North, and Judith Clarkson

University of Texas Press | Austin

Support for this book comes from an endowment for environmental studies made possible by generous contributions from Richard C. Bartlett, Susan Aspinall Block, and the National Endowment for the Humanities.

Requests for permission to reproduce material from this work should be sent to:
    Permissions
    University of Texas Press
    P.O. Box 7819
    Austin, TX 78713-7819
    www.utexas.edu/utpress/about/bpermission.html

♾ The paper used in this book meets the minimum requirements of
ANSI/NISO Z39.48-1992 (R1997) (Permanence of Paper).

Library of Congress Cataloging-in-Publication Data

The impact of global warming on Texas / edited by Jurgen Schmandt, Gerald R. North, and Judith Clarkson. — 2nd ed.
    p.   cm.
  Includes bibliographical references and index.
  ISBN 978-0-292-72330-6 (cl. : alk. paper)
  1. Global warming—Texas.   2. Climatic changes—Environmental aspects—
Texas.   I. Schmandt, Jurgen.   II. Clarkson, Judith.   III. North, Gerald R.
  QC981.8.G56147 2011
  363.738'74209764—dc22

                  2011006614

ISBN 978-0-292-73324-4 (E-book)

Contents

Foreword

Avoiding dangerous climate change will be one of the defining challenges for humanity in the twenty-first century. The Fourth Assessment Report of the Intergovernmental Panel on Climate Change (IPCC), published in 2007, documents compelling evidence that the climate is changing; human activities are the primary cause, and the consequences are serious. If we fail to act urgently and effectively to mitigate the causes and adapt to inevitable climate change, catastrophe looms for future generations of people, and the evolution of natural systems that support human society will change dramatically.

Climate experts agree that policymakers cannot wait until all the scientific questions are answered before they act, since by then it may be far too late to do so. We must approach climate change as a risk management problem. To do so requires three distinctly different but complementary types of responses.

First, scientists need to continue to work diligently to improve our understanding of how the climate system works, how it is likely to change in the future, how such changes may affect natural ecosystems and human society, and how we might adapt to these changes. Continuing scientific advances will be important to policymakers in the future as they deal with increasingly contentious and complex decisions on how to respond to the risks that a changing climate will pose.

Second, policymakers need to consider what measures can be taken now to reduce emissions of greenhouse gases, which are projected to be the primary drivers for future climate change. Scientists project that greenhouse gases need to be reduced 80 percent below current emissions to avoid catastrophic impacts. Such measures would not put a halt to future climate change (we are too late for that), but there is hope that they would buy some time for both ecosystems and societies to adapt to the changes that will occur. Although initial measures can be modest, they will need to be strengthened as scientific understanding improves and as the evidence for potentially dangerous changes in the climate becomes more convincing.

The third response strategy is to anticipate the changes in climate that we expect to be unavoidable and to prepare for these through adaptation measures. Such measures can increase our resilience to change and reduce threats of harm to people, infrastructure, and ecosystems.

Global warming, which drives climate change, is a global matter requiring unprecedented cooperation among nations. Moving forward recognizes that a future international agreement must meet certain key objectives. It must:

- have broader participation with fair goals, including all industrialized and key emerging economies;
- generate outcomes that will result in real progress on both mitigation and adaptation in all nations over the longer term;
- provide incentives to invest in developing countries and share transformative environmental technologies;
- maximize the deployment of existing climate-friendly technologies; and
- support the implementation of efficient carbon tax or market programs.

How can the individual American citizen influence the outcome of a global environmental issue that is already challenging the wisdom and resources of the world's governments and international agencies? The answer, simply put, is that it is the individual citizens who must create the environment of opinion that will encourage governments to act. And it is the individual citizens who can take actions themselves to reduce their personal emissions and who can support the policies that an effective response to the risks of climate change will demand.

The Impact of Global Warming on Texas provides a comprehensive and understandable assessment of what is currently known about the threats and opportunities posed by climate change in Texas. This book connects global climate change to the expected changes in local climates and impacts where we Texans live, work, and play. I am pleased to recommend this book. It will inform policy, enrich climate science education in our colleges and universities, and provide the basis for individual and public policy actions on both mitigation and adaptation.

Neal Lane
Malcolm Gillis University Professor
Senior Fellow, James A. Baker III Institute for Public Policy
Rice University
Houston, Texas

The Impact of Global Warming on Texas

Introduction
Bill Dawson

This book is being published at a moment of singular change and choice for Texans. As the authors make clear, the impact of global warming on the state is expected to take two basic and interrelated forms. One involves the effects of warming-caused climate change itself. The second involves the effects of any actions taken to reduce that warming.

First, geography means Texas will experience a challenging assortment of the climate changes that scientists have concluded are already spinning off from a man-made atmospheric warming trend. Chapter 10, for instance, includes a compelling, though hardly comprehensive, list of some of the things that scientists project for the state: "temperatures will rise, heat waves will occur more frequently, it will be drier west of the Interstate 35 corridor, severe weather will become more frequent, in-stream flows will fall, biodiversity will decline, and the sea level will rise."

Meanwhile, because of its energy-intensive economy and way of life, Texas will acutely feel the effects of any new policies designed to reduce emissions of carbon dioxide (CO_2), the principal greenhouse gas blamed for man-made warming. Carbon dioxide is emitted whenever fossil fuels, including oil and coal, are used. In Texas, the sources include millions of motor vehicles, sprawling oil and chemical complexes that serve much of the nation, and the coal-fired power plants that produce much of the state's electricity. Chapter 8 examines the factors behind Texas' high ranking among states emitting the most carbon dioxide. Two telling details: Texas leads the United States in overall energy use, with more than a tenth of the national total, and also in the consumption of coal.

This new, second edition of *The Impact of Global Warming on Texas* can aid in making the decisions that now confront the state. It is a completely revised version of the original edition, published in 1995. The authors—a distinguished team of climate scholars representing a variety of disciplines—present up-to-date interpretations of experts' current knowledge of the scientific, economic, and policy aspects of climate change in the state.

The book also offers recommendations for two major realms of possible response to global warming, reducing its impacts and adapting to those effects. The authors' work will nourish the understanding of policymakers and ordinary citizens alike at a particularly significant time in the climate issue's trajectory. In Chapter 10, Jurgen Schmandt notes that leaders now have "fewer excuses to defer action because of scientific uncertainty" than they did when the book's first edition was published, because researchers have resolved key questions in the ensuing 15 years. In addition, he writes, scientists have established that "climate change is no longer a distant possibility but is occurring now."

The likelihood that more-ambitious national policies will be adopted to reduce carbon dioxide emissions, including regulatory mandates, has increased as accumulating scientific findings and solidifying interpretations of that research have influenced and intersected with a number of other events and trends.

President Barack Obama promised change in his 2008 campaign for the White House, including a reversal of Texan George W. Bush's refusal to launch a regulatory attack on human-caused climate change. Within just a few days of taking office, Obama signaled that he would make good on that pledge when his administration proposed major increases fuel economy for cars and light trucks. The regulations, adopted in April 2010, include the nation's first-ever standards for greenhouse gas emissions.

Still to come, at this writing, is final action that would fulfill Obama's campaign pledge to enact a sweeping regulatory program aimed at reducing emissions of greenhouse gases from sources other than vehicles. Major steps have been taken toward that goal, both administratively and in Congress, but the issue is highly contentious and the outcome remains unclear. A major bill to limit greenhouse emissions narrowly passed the House in 2009, but the Senate has not approved that measure or its own climate legislation. On a separate track, the U.S. Environmental Protection Agency (EPA) in December 2009 finalized action on its formal conclusion that greenhouse gases endanger human health and the environment. This "endangerment" finding is a legal prerequisite before the EPA regulates such gases on its own initiative under the federal Clean Air Act. The finding is being challenged in court by a number of private interests and states, including Texas.

Obama's election, significant as it was for the climate issue, was just one in a series of major developments that have transformed the character of climate change as a political and economic issue in the last few years. Hurricanes Katrina and Rita, which devastated parts of Louisiana and Texas in 2005, profoundly heightened awareness and concern about the effects of global warming, including stronger storms, that climate scientists project. In 2006, a documentary film and book by former vice president Al Gore, both titled *An Inconvenient Truth,* used hurricane images among other visual and verbal tools to persuade the public to take climate change more seriously. Media coverage of a stream of scientific findings, including the 2007 reports of the Intergovernmental Panel on Climate Change (IPCC), likewise accelerated attention to the issue.

The IPCC, reflecting the work of hundreds of scientists around the world, declared that evidence of warming was "unequivocal" and that most of the warming that has occurred over the previous half-century was "very likely" the result of greenhouse emissions from human activities. The organization's reports contained warnings about possible consequences of deep concern for Texas, including hurricanes ("likely" to grow stronger, according to the IPCC), rising sea level (which boosts hurricane storm surges), and climate events including heat waves, high temperature extremes, and heavy precipitation ("very likely" to be more frequent). Early in 2007, *An Inconvenient Truth* won an Academy Award. Later in the year, Gore and the scientists who worked on the IPCC reports were jointly awarded the Nobel Peace Prize for heightening awareness of the issue.

In 2008, just four months before this edition's preview appearance in an online format, Hurricane Ike roared ashore in Texas, bringing destruction to Galveston and the Bolivar Peninsula, knocking out electric power to millions of Houston-area residents and exacting an enormous economic toll that made the storm one of the costliest in American history. In the 2008 presidential campaign, both Obama and his Republican rival, Senator John McCain of Arizona, endorsed regulatory proposals for battling climate change that had far more in common with one another than either did with the nonregulatory approach followed for eight years by Bush. By the end of Bush's second term, the national discussion about climate change, dominated by a clamorous debate over science until not that long ago, had become more focused on policy questions about what to do about this phenomenon that was being identified by a growing body of research.

Along with changed perceptions and a shifting public dialogue, climate change has increasingly blended with related concerns in the political arena, news coverage, corporate decision making, and elsewhere. The result is a larger, composite issue that encompasses global warming, energy policy, pollutants other than carbon dioxide that come from the same sources, dependence on foreign oil sources, national security, and the broad concept of sustainability. Terms like "going green" and "carbon footprint" quickly became widely familiar through repeated usage in the popular media, both news and entertainment. The *New York Times,* for instance, introduced "Green Inc.," a multi-reporter blog that focuses on the intersection of business, the economy, and the environment. The newspaper described the blog's mission this way: "How will the pressures of climate change, limited fossil fuel resources and the mainstreaming of 'green' consciousness reshape society? Follow the money."

When the economic meltdown that took place during the latter part of 2008 morphed into a deepening recession, it focused new attention on concepts like "green stimulus" and "green New Deal." They refer to government programs designed to revive the economy through spending to boost alternative energy, energy conservation, energy efficiency, and the like. Obama endorsed such ideas in his campaign, and his economic stimulus proposal included funding for such initiatives.

All these changes, then, have contributed to a growing array of choices for Texans about how they will respond to the interconnected issues of climate, energy, and economy. At the state government level, for instance, Bush's decision not to pursue regulations to limit global warming created a policy void that has increasingly been filled by the actions of individual states and multistate compacts to tackle the climate issue directly. Texas has not been one of them. "So far, Texas has done very little to address the problem of global warming," Judith Clarkson writes in Chapter 8. "In fact," she adds, "the official policy appears to be to wait and see what the federal government does."

With the Obama administration adopting an approach to climate change very different from that followed by Bush, a key question looms larger than ever: Will government leaders in Texas still wait to see what federal policy means for the state (while trying to influence that policy, to the extent they can), or will they also begin to shape a complementary Texas response to global warming?

Early in 2009, as the new national administration was just starting to implement its own policies, it seemed that state lawmakers might begin forging a new path for Texas. The new speaker of the House of Representatives, Republican Joe Straus of San Antonio, was an outspoken proponent of energy efficiency and had told *Texas Monthly* magazine that he looked forward to the day "when alternative energy is a mainstream source to power Texans' lives." Other indications also gave advocates strong hope that the 2009 legislative session would produce various bills that would have the effect of cutting greenhouse emissions.

That optimism was largely unfulfilled, however, as major bills on renewable energy, energy efficiency, and related matters that had seemed likely to pass instead died in the session's last hours in a legislative logjam. Still, other bills were passed that for the first time in Texas explicitly recognized climate change attributed to greenhouse emissions with requirements for state officials to carry out certain planning and information-collecting activities.

One of these successful measures, called "No Regrets," could lead to future legislative action. The proposal, by an Austin Democrat, State Senator Kirk Watson, had failed to pass in the 2007 session. As enacted in 2009, it required state officials to solicit and analyze strategies for cutting greenhouse emissions that would result in net savings for consumers or businesses, bring no financial costs, or help Texas businesses "maintain global competitiveness." The State Energy Conservation Office was to submit a report on the strategies to the Legislature in time for possible consideration in the 2011 session.

In Chapter 10, Schmandt argues that Texas has been pursuing what is essentially "a hidden climate change policy," motivated largely by concerns about energy independence and efficiency. The hidden policy, he asserts, includes various efforts to advance alternative energy and energy efficiency and is more ambitious than many people recognized. Schmandt calls instead for "a comprehensive policy that links climate change to energy independence, regional security, and the management of natural resources," along with the establishment of a state Office of Energy, Security, and Climate to focus such a unified, overt commitment.

Of course, taking such actions would require an acknowledgment that man-made climate change is an issue worth major and explicitly directed attention and resources. There have been suggestions that some high-ranking state officials do not share that viewpoint. For example, an advisory panel appointed by Governor Rick Perry and made up of top officials of three state agencies argued in late 2008 against federal regulation of greenhouse emissions by the EPA. The panel's report signaled doubt about the interpretations of mainstream climate science in a brief passage asserting that "recent climate research calls into question prevailing public perceptions of the cause and extent of global warming."

High-level skepticism about the conclusions of mainstream climate science was expressed more explicitly and forcefully in February 2010, when Texas Attorney General Greg Abbott filed the state's petition challenging the EPA's endangerment finding about greenhouse gases. At that time, Abbott and Perry leveled harsh criticism at the IPCC and its conclusions. They based their arguments on news reports about a few errors in the IPCC's voluminous 2007 reports and about previously private e-mails of some climate scientists, which had been leaked or stolen from a British university and which climate change skeptics alleged were indications of data manipulation. Abbott said the IPCC was "not trustworthy," and Perry's office said it was "discredited."

Leading climate scientists at Texas universities soon disputed those arguments. Kenneth P. Bowman, who heads Texas A&M University's Department of Atmospheric Sciences, issued a statement saying the department's entire faculty believed the EPA finding was based on "good science." A few weeks later, scientists from Rice University, Texas A&M (including Gerald North, one of this book's editors), Texas Tech University, and the University of Texas at Austin coauthored an op-ed column in the *Houston Chronicle*, declaring that "the science of climate change is strong." (Before the state's petition was filed, the Associated Press conducted an "exhaustive investigation" of the disclosed e-mails, concluding that they "don't support claims that the science of global warming was faked." Subsequently, the British House of Commons Science and Technology Committee reached a similar conclusion.)

Perry has consistently argued that federal regulation of greenhouse gases would cripple the Texas economy. A better approach, he has said, is to make alternative energy technologies, such as wind-generated energy, less expensive, along with fostering investment in nuclear power and in technologies for capturing and sequestering carbon emissions.

The growth of Texas' wind-power industry is an instructive case study in the complexity and ambiguity of the stances and actions taken by state officials regarding human-caused global warming. Perry, for one, has delivered what some might regard as mixed messages. For instance, the same press release from the governor's office that called the IPCC "discredited" also suggested that cutting greenhouse gases is a worthy goal. It noted that Texas has "reduced carbon dioxide emissions more than nearly every other state," has done so without "government mandates or extravagant fines," and has "installed more wind power than any other state." Similarly, when Perry announced a $10 billion private-public partnership to expand wind-energy infrastructure in Texas in 2008, his office declared that "for every 1,000 megawatts generated by new wind sources, Texas will reduce carbon dioxide emissions by six million tons over the next 20 years."

On some occasions, Perry has made light of concerns about human alteration of the climate with greenhouse emissions. In 2007, news reports said he remarked to an audience in California (site of the most dramatic state-level actions to limit greenhouse gases), "I've heard Al Gore talk about man-made global warming so much that I'm starting to think that his mouth is the leading source of all that supposedly deadly carbon dioxide." In June 2009, voicing concern that the EPA might declare carbon dioxide (and other greenhouse gases) an air pollutant deserving regulation, he told reporters in Austin, "I mean the idea that CO_2 is a toxic substance is a bit hard for this, you know, agricultural scientist to get his arms around when Nobel Peace Prize or Nobel laureates have talked about CO_2 in a very positive sense, when you talk about the Green Revolution," which boosted agricultural production in developing nations.

Such pronouncements raise an obvious question: Why announce that the state's reductions in carbon dioxide are a benefit of the wind industry's growth in Texas—or why mention those reductions at all in a positive way—if the gas is only a "supposedly" harmful substance and the leading scientific body saying that it is harmful is "discredited"?

In any event, Texas now produces more electricity from wind generation than any other state, thanks in large part to actions by the Legislature in 1999, and again in 2005, requiring utilities to produce a certain percentage of electricity from renewable sources. In 2008, the Public Utility Commission launched a $5 billion project to increase transmission capacity, then in early 2009 pushed it ahead by giving seven utilities assignments to construct the new electric lines to carry wind energy from West Texas to North Texas and the Houston area.

Also in 2008, the prominent Texas businessman T. Boone Pickens gained considerable national attention through the heavy promotion of his Pickens Plan, a proposal to help wean the United States from its dependence on foreign oil through the use of wind and solar power and natural gas. Initially, Pickens suggested substituting natural gas for imported oil in some vehicles, including private cars, while replacing electricity now generated by that natural gas with more wind power. By early 2010, he had modified his plan to de-emphasize wind power and personal vehicles and emphasize converting large, diesel-burning 18-wheelers to use natural gas. Pickens stresses the benefits for national security of reducing reliance on foreign oil, rather than the lowered greenhouse emissions that would result from replacing oil with carbon-free wind and lower-CO_2 natural gas. He told an audience at Rice University early in 2009 that "global warming is Page 2 for me." Nevertheless, he has made common cause with campaigners for attacking climate change, such as Al Gore and the Sierra Club.

Such alliances might seem odd (Gore is a liberal Democrat and Pickens is a conservative Republican), but they typify the synergistic opportunities that have arisen in recent years, along with the blending of the climate issue with other concerns. Texas is rich in natural gas reserves, and the fuel was being mentioned by some Texas officials as far back as the late 1980s as a weapon against global warming.

The promotion of natural gas and wind power in the Pickens Plan is only one recent sign of a growing recognition among private-sector leaders in Texas of a changing political and economic landscape with regard to the climate issue. For example, several years ago, before retiring as president of Shell Oil, John Hofmeister began calling for a "culture of conservation," including national policies that recognize global warming as a problem requiring action. In 2008, commenting on the possible EPA regulation of greenhouse gases that Perry opposed, officials of San Antonio-based AT&T said they were ready to help the federal agency cut emissions through the development of next-generation information and communication technologies.

Probably no corporation has been more closely watched or more strongly criticized on the climate issue than the Texas-based giant Exxon Mobil. During the George W. Bush administration, Exxon came under increasing attack for its influential opposition to the regulation of greenhouse gases. In 2006, the Royal Society, Britain's national sciences academy, called on the company to stop supporting groups that had "misrepresented the science of climate change by outright denial of the evidence" and for Exxon itself to cease issuing "inaccurate and misleading" statements on climate science. Exxon still has its critics over the climate issue, to be sure, but there is no denying that its public pronouncements have shifted dramatically. In 1991, the *New York Times* reported that Exxon's chairman, making a speech in Houston, had "expressed doubt that theories on global warming would eventually prove accurate."

In late 2008, an article in the *Times* included this passage:

> Gingerly, over the last three years, Exxon has moved away from its extreme position (on global warming). It stopped financing climate skeptics this year, and has sought to soften its image with a $100 million advertising campaign featuring real company executives, scientists and managers. One of the ads said the company aimed to provide energy "with dramatically lower CO_2 emissions."

Then, early in 2009, the *Wall Street Journal* reported this development: "The chief executive of Exxon Mobil Corp. for the first time called on Congress to enact a tax on greenhouse-gas emissions in order to fight global warming."

Climate-conscious transitions are evident in Texas beyond the arena of corporate policy. Sustainability initiatives have gotten under way at a number of colleges and universities, as administrators and students embrace policies and goals related to energy conservation, carbon reduction, and related matters. At the University of Texas, for example, a sustainability plan prompted by a student proposal was announced in 2008. The university said it would "integrate sustainability in academic programs, operations, campus planning, administration and outreach" in one of the most rigorous initiatives in the state. In early 2010, students at both the University of Texas and Texas A&M voted to levy small fees on themselves to fund campus sustainability projects. Meanwhile, curriculum offerings at various institutions of higher education reflect the same trend in different ways. In one example (of many that could be cited), Midland and Odessa colleges in West Texas recently announced that they would offer instruction in wind-energy technology.

City government is another arena where programs and policies that implicitly or directly address climate change have multiplied in Texas, reflecting a municipal trend that has paralleled state government initiatives across the country. Houston, Dallas, and other cities, for instance, formed the Texas Clean Air Cities Coalition, which joined forces with environmentalists and others in 2006 to oppose a proposal by the Dallas-based utility company then called TXU to build 11 new coal-fired power plants in Texas. The opposition was based on concerns about the health effects of smog-forming pollutants, as well as the carbon dioxide that the power plants would release over decades of operation. In 2007, the company was acquired by buyers who, after consulting with opponents of the plants, dropped plans for all but three of them. In 2008, the coalition, by that point numbering 37 cities representing more than half the population of Texas, and the Environmental Defense Fund dropped their joint opposition to the expansion of an NRG power plant after the company agreed to implement measures that would offset a large part of the coal-fired plant's carbon dioxide emissions.

By early 2010, 31 Texas mayors had committed their cities to the goals of the U.S. Mayors Climate Protection Agreement, a pact managed by the U.S. Conference of Mayors and signed by more than 900 mayors. Texas participants range from large cities (including Austin, Dallas, Fort Worth, and San Antonio) to smaller ones like College Station and Sugar Land. The agreement commits participants to try to meet or exceed the carbon dioxide reduction targets in the Kyoto Protocol, an international climate treaty, and to urge their states to do the same. The agreement is discussed in Chapter 8, which also provides a detailed account of climate-related initiatives in Austin. The capital city, Judith Clarkson reports, takes pride in being a municipal leader in reducing fossil fuel consumption.

Participation in the U.S. mayors pact does not convey the full extent of the efforts by Texas cities to cut greenhouse emissions and otherwise deal with climate change. Houston, the state's largest city, has not signed the agreement but has launched a number of climate-linked projects in recent years. They include a project undertaken in concert with the Clinton Climate Initiative (named for the former president) to improve energy efficiency in municipal buildings through retrofits.

Besides such actions aimed at reducing carbon emissions, Houston has taken steps to be better prepared for hurricanes following its experience with Hurricane Ike in September 2008. Two months later, responding to the massive power outages that Ike caused, Houston's mayor, Bill White, appointed the Task Force on Electric Reliability, made up of individuals with relevant expertise, to recommend how to make the power grid "more durable and resilient."

Heightened preparedness for strong hurricanes like Ike also has been on the minds of Houston-area citizens, as reader comments on the *Houston Press* website in early 2009 illustrated. Regarding an article relating Galveston's struggles to recover from the storm, a Galveston resident wrote, "The No. 1 issue is now to minimize or eliminate catastrophes in the future, whether it involves raising houses or providing surge protection on the bay." A Houstonian added this comment, referring to the famous seawall that Galveston residents constructed after their city was devastated by the catastrophic hurricane of 1900: "Galveston has a dike, but it stops after ten miles. Did Ike know it was not supposed to go beyond the dike and flood the city from the back side? I grew up in Holland, where taxpayers spend billions of dollars on keeping their feet dry. What do we do here? Pray!"

Since then, government officials and others have begun considering the construction of a massive Ike Dike, a Netherlands-inspired system of levees, seawalls, and floodgates to extend the storm-surge protection of Galveston's seawall in both directions along the coast. As envisioned by William J. Merrell, the George P. Mitchell Professor of Marine Science at Texas A&M University at Galveston, a wall 17 feet high would extend 60 miles, from High Island west to San Luis Pass. At this writing, the county judges of six counties in the Houston-Galveston region are planning a formal study of the concept.

Preparedness for stronger hurricanes and other possible outcomes of climate change is a key theme that runs through the diverse contributions assembled in *The Impact of Global Warming on Texas*. Neal Lane, in the Foreword, urges that global warming be treated as "a risk management problem." Time will tell whether Texans and their leaders adopt that viewpoint. If they do, the careful, comprehensive, and evidence-based assessments in this book can serve as a useful guide. In any event, the book's publication means the state is being briefed at a crucial time about the risks and opportunities that some of Texas' most knowledgeable climate experts foresee.

THE AUTHORS' SUMMARY OF THEIR KEY CONCLUSIONS

Climate Science and Climate Change: Climate science has evolved over the last 35 years to a point where predictions by climate models can be considered to have significant information content. The greenhouse effect has been clearly established as a driver of climate change, and the main agent is the continuing increase in the concentration of carbon dioxide in the atmosphere. There are several ways of assessing the status of climate change research. The most recent and comprehensive assessment is from the Intergovernmental Panel on Climate Change, the Fourth Assessment Report, released in 2007. According to this report, greenhouse gases are expected to cause global temperatures to rise 5.4°F (plus or minus 1.8°F) by the end of the century. Temperature changes in Texas are expected to be comparable. A notable feature of the predictions is the expansion of the tropical zone, familiar in summer for Texans, to include more of the spring and fall. This could lead to less rainfall, especially in regions that are already dry. Other important effects include possible changes in El Niño (climate variability) and hurricane behavior; further research will more accurately specify these and other effects.

The Changing Climate of Texas: Texas precipitation increases dramatically from west to east. The seasonal patterns of precipitation also vary greatly across the state (e.g., dry winters in the west, more even distribution seasonally in the east). Texas experiences a variety of severe weather, such as tropical storms, tornadoes, drought, and flooding. The wide variations in weather and climate across Texas imply a broad range of vulnerabilities to climate change. Averaged across Texas, the temperature has been increasing over the last few decades. Precipitation has also steadily increased over the past century, but with variation among regions. In the future, Texas temperatures are likely to continue rising. Precipitation changes are much less clear, with most models projecting a decrease. Even if precipitation were to remain stable, rising temperatures would increase evaporation and dryness. The expected changes in temperature and precipitation will have an impact on other sectors of the state's resources.

Water Resources and Water Supply: Taking water flows to the coast as a measure of river basin impact, we calculate which changes will occur by mid-century under constant and changing climate conditions. Considering only population growth and the resulting increased water demand, flows will be reduced by about 25 percent under normal conditions and by 42 percent under drought conditions. When also considering climate change (3.6°F increase in air temperature and 5 percent decrease in precipitation), 2050 projected flows to the coast are 70 percent of the 2000 values under normal conditions and 15 percent under drought conditions.

Coastal Impacts: Two direct effects are already observable in the instrumental record: rapid sea-level rise and rising sea temperature. The rates of sea-level rise are especially high in Texas because of the added effect of land subsidence, which is caused by oil and groundwater extraction. The increasing temperatures are already manifesting indirect changes in habitats and water quality.

Biodiversity: Climate is a key determinant of species distribution. As the earth warms, species tend to shift to northern latitudes and higher altitudes. But climate change represents just one of a set of stressors. Other changes challenging fauna and flora are due to land development, habitat fragmentation, invasive species, chemical stressors, and direct exploitation. Comprehensive assessments in each of the state's ecological regions (coastal marshes, forests, deserts, prairies, and western mountains) are needed to develop science-based management practices for wildlife and plant communities.

Agriculture: Agriculture in the United States and Texas is sensitive in terms of land and water uses, as well as crop and livestock production. In terms of agricultural-based economic welfare, however, the simulated effects of climate change are not large. We find that under the climate change conditions simulated herein, statewide Texas cropped acreage declines by about 20 percent.

Urban Areas: Coastal population centers, from Houston to the Lower Rio Grande Valley, are vulnerable to rising sea level, increased storm intensity, and the accompanying flooding. All major Texas cities face the possibility of impacts on air quality, energy, health, and other temperature-related effects. All major cities face the prospect of declining water resources within the time frame examined here.

Greenhouse Gas Emissions: Because of its large population and energy-intensive economy, Texas leads the nation in oil refining capacity and energy consumption, accounting for more than one tenth of total U.S. energy use and 11 percent of U.S. greenhouse gas emissions. Although more than 30 states have taken some measures to address the issue of global climate change, Texas has not been willing to take direct action. It has, however, been a leader in renewable energy, and following legislative action in 1999 and 2005, Texas leads the nation in wind power production. There are many other, cost-effective measures that could be taken that would both reduce greenhouse gas emissions and improve the competitiveness of Texas products.

Economy: Looking to mid-century, it is clear that the cost to Texas of a national cap-and-trade policy would likely exceed any possible measurable benefit in terms of avoided damages. Over a longer time frame, if the harmful impacts of climate damage continue to increase, the cost-benefit balance might shift, but time is not on our side. Texas would benefit economically by taking stronger actions today to address climate change impacts at the state level and by supporting the adoption of cost-effective, equitable policies at the national level to limit greenhouse gas emissions and encourage the use of alternatives to fossil fuels.

Policy: Texas is a leader in the gradual shift to renewable energy. Energy and water conservation are also priorities, mostly at the community level. The driving forces of these policy initiatives are energy efficiency, resource conservation, and the income and jobs associated with industries developing alternative energy sources. These measures help to reduce greenhouse gas emissions. Thirty states have joined regional climate change alliances. Texas has not done so. We recommend that Texas develop a comprehensive climate change policy to serve the goals of reducing greenhouse gas emissions, increasing energy independence, ensuring regional security, and improving management of water, air, land, and wildlife.

CHAPTER 1

Climate Science and Climate Change
Gerald R. North

This is the second edition of a book assembled by the same editors in 1995. The first edition was one of the earliest attempts at a regional assessment of the impact of climate change; in our case, the application was to the Texas region. There has been significant progress in climate research since then, and the present publication is our response to the many requests to revise and update the book. Our aim then and now is to provide an objective assessment of the impacts of climate change on the Texas region. Our target audience is the well-educated layman, especially those in a policymaking role, whether governmental or private. Our object is not to scare but rather to let our readers know what the current thinking is in the field so that they can plan for changes that are likely to occur.

Separate chapters deal with different aspects of the problem, and each is written by an expert or experts in the field. The chapter authors come from a number of institutions in and outside Texas, and it was our intention not to represent any particular point of view. The book contains very little original research, but rather it represents, to the best of our ability, the state of the art of the science at this time as drawn from the published literature on the subject.

The current chapter deals with the science of climate change, including the question of whether it even *is* a scientific undertaking, its foundations, paradigm changes in its history, and how we go about assessing the state of the art of climate science and the potential impacts of climate change. Finally, some results are presented along with a few implications for the future climate of the Texas region.

MODERN CLIMATE SCIENCE

Climate science as we know it took a dramatic turn in the 1970s. Before then, climatology consisted of the collection and classification of data taken from conventional sources from all over the world on all the possible time scales, such as annual, monthly, and even daily tabulations of the summary statistics of weather at particular locations. Often the statistics were averaged over regions, even to continental scale. Climatology books and courses provided diagrams and maps that helped classify climates based on factors such as soil type, temperature, and moisture availability. The diagrams and maps were accompanied by qualitative discussions of the climate of a particular region in the context of its topography, geography, latitude, seasonality, and other parameters. Some quantitative measurements were also being conducted in such areas as surface energy balances between radiation heating, conduction of heat into the atmosphere, and the latent heat exchanges from evaporation. For the most part, global and regional climates were regarded as stable, and such concepts as "drought of record" were valuable planning tools. Some primitive models of global climate were finding their way into the literature, such as those of Budyko (1968) and Sellers (1969).

Numerical models for forecasting weather have been around since the 1950s, when the first operational digital forecasts were carried out at the Institute for Advanced Study in Princeton, New Jersey. As computers became more powerful and the software evolved, numerical models were gradually adopted as one of the tools used by weather bureaus around the world to forecast weather, although the human forecaster still did most of the work. The discipline imposed by forecasts and having to face the music day in and day out caused the gradual evolution of the weather models, until their utility was undeniable in the early 1970s. It was a natural step to adapt these models to address the problem of climate simulation.

Unlike econometric and other models that are sometimes based on balancing flows of quantities, but with liberal helpings of regression coefficients, weather forecasting models are physical models depending on the laws of fluid mechanics, electromagnetic radiation, and thermodynamics. The physics places strong constraints on the model solutions. A weather model forecast is analogous to computing the trajectory of an ideal projectile based on Newton's laws of motion. If the projectile's initial position and velocity are known, the parabolic trajectory is determined as shown in every high school physics book. Picture, for example, a dot where the projectile is computed to land. A small random error in the initial position or velocity will lead to a small error in the location of the final dot on the earth's surface. If we repeat this procedure over and over with a different error each time, we will obtain a swarm of dots at the end point (Fig. 1.1). The longer the trajectory, the larger the spread of dots, as the trajectories diverge more with time of flight.

The weather forecast is similar to the single trajectory computation, except we must follow the millions of air parcels simultaneously along their paths. The problem is further complicated by the variety of forces that the

FIGURE 1.1.   A distribution of 1,000 landing locations (an ensemble) for a shot where knowledge of the location of the gun is in random error before each discharge. The error is bivariate normally distributed, with the standard deviation in the *x*-direction 1.5 times that in the *y*-direction. Each shot can be thought of as a weather event, but the mean and the dimensions of the swarm represent the climate. A shift of the mean or dimensions of the pattern is the analog of climate change.

parcels encounter along the way, not least of which are the forces exerted on one from the others. So instead of the back-of-the-envelope calculation of a single projectile's end point, we have to compute the end points of the trajectories of millions of air parcels from their original positions (pressure distribution) and velocities (winds) on a rotating sphere with nonuniform temperatures (buoyancies). The planet is heated more near the equator than at its poles, which causes a planetary convection of air called the general circulation. Also, heat is released as moist parcels form droplets that make up clouds, and this gives them additional buoyancy jolts (formation of the droplets releases large quantities of heat, the opposite of the evaporation that cools our skin). Despite all these complications, the computations go pretty much like the single projectile case, except that we must keep track of millions of parcels, and that requires very large and fast computers. In a modern global forecast model, the inevitable errors in initial location and velocities lead to growing errors that become intolerable after a few days, but the erroneous evolution of the simulated weather beyond a few days still looks like real weather (the statistics of temperature, storm passages, and so on look similar to what actually happens). The forecast fails to predict correctly the weather on a particular day at a particular location. Nevertheless, we still have the progression of midlatitude storms from west to east, and the seasonal cycle looks remarkably faithful to observations.

What would climate change look like in our projectile model? An analogy to climate change is the shift of some external parameter such as the gravitational acceleration (nominally $g = 9.81$ meters per second^2). If we change the value to, say, 12 meters per second^2, our initial, error-induced swarm of end points will now occupy an elliptical pattern on the surface, but they will be somewhat closer to the initial projection point. The entire pattern shifts, and its shape might change as well.

When we run our weather model as a climate simulator, we often want to know how the climate changes as we change some parameter in the model, such as the concentration of a gas in the atmosphere or perhaps the solar brightness. We are interested in the new summaries of weather statistics (or the projected climate). The toy projectile provides a simplified explanation of the way climate simulations are conducted. If the change in the acceleration of gravity for the toy projectile model is small, the swarms will overlap, but even small changes might be detected if enough data are collected, and such small changes in the statistics could be important. Changing the value of one of the external parameters, such as the gravitational acceleration, in the toy model is similar to a change in climate induced by, say, a change in the sun's brightness or any of the other possible parameters that might cause a change in the weather statistics (climate).

Climate models are more complicated than ordinary weather forecasting models, since they need to include additional processes that are not very important in the weather forecast. This is because even slow, tiny changes in, say, solar brightness will cause the statistics to shift to new values. The main new features that must be included are comprehensive treatments of the details of radiation passing through the atmosphere (e.g., absorbing, scattering, emitting), water vapor transport, cloud physics, and ground surface properties. Moreover, long-term changes in the ocean play a big role in climate change but hardly any in short-term weather forecasts.

The success of weather forecasting led to a worldwide system of data collection and assimilation. These observations, taken from the ground and from balloons, could be archived after use in the forecast initialization and made available for climatological research. Another breakthrough occurred in the mid-1970s, when satellites launched weather observation into Earth orbit. In the beginning, these observations were intended mainly for initializing weather forecasts, but it was quickly noted that these orbiting observatories could collect vital climate data as well. Such elusive observations as cloud cover, sea ice cover, and area averages of precipitation (especially over the oceans) are now routinely observable on a global basis. Satellite data can be used to check climate model simulations of the present climate. Data routinely collected by ships passing over the world's oceans reveal not just the surface parameters but those below as well. All of these observations are being collected and sorted into useful data sets for global climate research. The synergy of modeling and global observing systems has continued through the decades since the mid-1970s.

So what is the climate change that has been the cause of all the concern? Even the simplest global climate model, which balances the rate of visible sunlight being absorbed by our planet and the rate of energy emitted from Earth to space, predicts a higher surface temperature if the planet has an atmosphere with some opacity in the infrared portions of the electromagnetic spectrum. In fact, our planet would scarcely be warm enough for habitation by our species if there were no greenhouse effect (mostly due to water vapor, which is a greenhouse gas). That model is a back-of-the-envelope calculation that has been known for a century, at least qualitatively. A paper published by Gilbert Plass (1956) estimated that doubling the carbon dioxide in the atmosphere would increase the planet's temperature by about 2.5°C (4.5°F), a change very close to the currently accepted value. Plass spent the last years of his career as a professor of physics at Texas A&M University, but he left this research area because he thought his theory could not be improved without more accurate measurements of the absorptivity of greenhouse gases to infrared radiation. Now, 50 years later, that task has been essentially completed.

The problem is that we are increasing the infrared opacity of our atmosphere, meaning that gases that fully absorb some wavelengths in the infrared spectrum are accumulating and trapping energy in the form of heat. Rough computations with simple models can give us approximate measures of climate change due to changes in greenhouse gases (plus or minus 75 percent), but a proper treatment of the feedback in the climate system requires much more comprehensive models. Moreover, estimations of global average temperature are not enough to help decision makers. Changes in precipitation and regional peculiarities of the warming cannot be computed with the simple energy balance models. For that level of detail, we must turn to the giant, comprehensive climate models that have grown out of weather forecast models, models that push the world's largest and fastest computers to their choking points.

Over the last three decades, data sets have been compiled that densely cover spatial areas and time frames. They close many gaps in our understanding of climate change, both natural change and that forced by changes in external parameters. We have comprehensive estimates of the history of atmospheric and oceanic temperatures and winds, currents, and other factors, and also of the agents that force climate to change, such as the opacity of the atmosphere following a volcanic eruption, and long records of the brightness of the sun taken from satellites outside the atmosphere, the concentration of greenhouse gases, the properties and composition of tiny particles afloat in the air, and changes in acidity of the oceans, to mention just a few. These measurements of causes and effects over the last three decades allow us to separate natural variability cleanly from the trends that are forced in the climate system. As we go back in time, the comprehensiveness and accuracy of the data degrade, so it is more difficult to draw the same rigorous inferences. Even with these limitations, however, the evolution of climate over geologic time down to the present is beginning to fit a consistent paradigm of climate change and its causes.

MODELS AND DATA

Although we may accept the paradigm that we can implement the laws of physics to simulate climate, via sophisticated numerical analysis on the world's best computers, there are always problems in the implementation. The simulation is accomplished by casting the problem onto a grid or similar construct covering the planet. Presumably, the finer the grid, the more faithfully are the equations from physics represented in the simulation. The coarseness of the grid spacing is a major limitation of the approach, but steady improvements in this area have been made (Fig. 1.2). Even so, clouds and precipitation patterns, particularly in the tropics, are much smaller than the grid boxes in today's models. Rather than try to simulate a thunderstorm 110 kilometers wide, a model must specify the cumulative effect of the various thunderstorms that would occur within an individual grid box. These fudge factors are adjustable parameters that inevitably enter the model making, and they will never exactly mimic the behavior of the real processes that they replace. Reducing the errors that arise from such approximation is among the most challenging problems in the field and is the subject of continual study.

As with the models, there are inevitable problems with the data. We would love to have data on temperature, precipitation, pressure, cloudiness, humidity, and all the other factors over a global range, mapped at fine resolution in both the horizontal and vertical dimensions. We would love to have centuries of it, but we cannot go back in history and make additional observations. The records we have often do not cover enough time and their spatial scale is not fine enough. Instead, we must get by with what we have and check our model simulations of past and current climate against the available data. In some cases we need to adjust those pesky parameters a bit to nudge the simulation to get a better fit with the facts (which are also approximate). Besides the global coverage of data, we need to augment our observational picture with field programs that help us understand individual

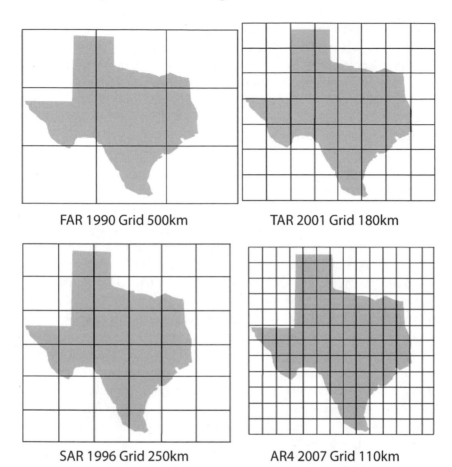

FAR 1990 Grid 500km   TAR 2001 Grid 180km

SAR 1996 Grid 250km   AR4 2007 Grid 110km

FIGURE 1.2.   The progressively finer grids used for climate model simulations in past IPCC assessments, from the First (FAR), Second (SAR), Third (TAR), and Fourth (AR4) Assessment Reports. Horizontal grid sizes are given in kilometers. Grids are superimposed on an outline of Texas for comparison only; the position of the map does not necessarily reflect exact locations.

processes such as cloud formation and soil drying. These programs are labor-intensive, often requiring multiple aircraft and delicate instruments, and of course they are limited to small regions of the planet. Field programs are essential for understanding small-scale processes (covering tens of miles) and for serving as ground truth for the satellites.

Aside from these concerns, there is the need to analyze the data and the output of the models. This analysis and testing form a large part of the climate change enterprise.

IS CLIMATE SCIENCE REALLY SCIENCE?

The question arises from time to time, especially among skeptics of climate change, "Is climate science really science?" In other words, is climate science something like astrology? Is the theory so qualitative, flexible, and tentative that it can accommodate or explain away any conceivable, new, seemingly contradictory data within its framework? Successful scientific theories should run some kind of risk or they quickly become as uninteresting as armchair prattle. Philosopher Karl Popper would say the theory should be "falsifiable," it should make predictions that if proven false invalidate the theory. Most philosophers of science now believe that such a strict criterion in modern science is too stringent because of the many assumptions that must underlie both the observations and the reduction of the theory to a specific test (Worrall 2003). This certainly appears to be the case in climate science, where the implementation of the laws of physics on a finite numerical grid and the derivation and interpretation of such data as those retrieved from instruments on satellites all involve many assumptions and approximations that are not always easy to justify.

Some people believe concerns about global warming are politically driven or even that they are the product of a sect of environmentalists analogous to a religious movement. Although there has been political passion (for example, a former vice president has been a visible and outspoken advocate of climate change concerns), the scientific consensus is hard to deny. At last, partisan political bickering about anthropogenic climate change is subsiding. There has also been a false impression that many scientists are opposed to the theory, but objective surveys of scientists who have recently published in mainline journals on the subject show overwhelming support for the proposition that human activities are responsible for the recent warming (personal communication, 2008, Prof. Arnold Vedlitz, Bush School, Texas A&M University).

The philosopher and science historian Thomas Kuhn, in his influential book *The Structure of Scientific Revolutions* (1970), describes the way science changes course from time to time. Typically science moves along at a steady pace following a certain paradigm, or way of thinking, until the current theory or way of describing things breaks down or comes up against an array of observations or internal contradictions. At this point some of its former adherents begin to cast about for a new way of approaching the problem. At other times, the paradigm might give in to stagnation. For example, there might be no new data available for generating new problems to solve, and scientists must have problems to solve; otherwise, they move to a new area where there are interesting things to do. The 1970s presented such a barrier for climate science. Rapid advances in technology awakened climate research from a state of stagnation. The original climate models were not invented to address global warming, but rather they were concerned with explaining past climates (for which new data were being analyzed and interpreted for such phenomena as the continental glaciations). As climate models and the accompanying tech-heavy data advanced (from, e.g., ocean floor drilling and retrieval of cores that could be used for inferring past climates), so did climate science. We moved into a phase that Kuhn would refer to as "normal" science, a period of slow but steady advance, as lots of scientific questions arise that can be addressed in the new framework. The framework was successful in answering those questions, and science hungered for more. Science and scientists have continued along these lines, answering question after question, puzzle after puzzle. The list of questions answered and understood through climate modeling would fill a book far thicker than the present one. This program will continue until another impasse is reached. (No such impediment is in sight at the present.) Although the relevant laws of physics are pretty well understood, there is plenty of work to be done in implementing and approximating those laws on computers. Likewise, understanding and expanding the observational database present many challenges, but this process is healthy and moving forward. So climate science fits the usual criteria of a newly emerging field in the scientific enterprise. There is no reason to think we are on the wrong path.

It is not always easy to separate scientific findings (so-called positive statements) from value judgments (normative statements); these latter include finger pointing, what ought to be done, who is to blame, and who pays (Dessler and Parson 2010). The normative issues are truly important, but they have little to do with the positive statements. If we can keep these two categories separated, we will make much greater progress in dealing with our problem.

A FEW QUESTIONS OFTEN POSED BY SKEPTICS

At seminars or during call-in sessions about climate change, several questions are brought up repeatedly by laymen or engineers and scientists from other disciplines. A few, with answers, follow:

1. Since water vapor is the strongest greenhouse gas and it is naturally occurring, why do climate scientists always pick on carbon dioxide and the other anthropogenically produced greenhouse

gases? *Answer:* Water vapor is indeed a very strong greenhouse gas, so much so that without water vapor (and perhaps clouds) in the atmosphere the planet would be about 60°F colder than it is. In addition, the amount of water vapor increases as the climate warms. Hence, it is a very powerful positive feedback mechanism, roughly doubling the response that we would expect normally from doubling carbon dioxide (expected to be about 1.0°C, or 1.8°F). Climate models are nearly unanimous on this point. But because the amount of atmospheric water vapor is so directly connected to global temperatures, there is nothing practical we can do to alter it substantially, other than keeping global temperatures in check by other means.

2. The concentration of carbon dioxide is only 380 parts per million (it makes up only 0.038 percent of the molecules in the atmosphere) whereas there is about 100 times more water vapor. How can adding a few more molecules of carbon dioxide matter? *Answer:* The air is pretty cold and dry where the emission of infrared radiation to space takes place. The number of carbon dioxide molecules is as much as a third the number of water molecules at that altitude (about 10 kilometers, or 6 miles). So the two are comparable. Each absorbs and emits in its own characteristic band in the infrared, and in important bands they do not overlap. Hence, the emissions and absorptions of these two gases both matter a great deal.

3. During the major glacial advances in the last million years, the temperature swings appear to lead the carbon dioxide swings by a few hundred years. How could carbon dioxide be the cause? *Answer:* The consensus among climate scientists is that carbon dioxide and methane are positive but very slow feedbacks in the climate system. It is the changes in the orbital elements of Earth's motion about the sun that trigger the glaciations (the Milankovitch effect). Once the glaciers begin to grow from this trigger, the concentrations of several important greenhouse gases are drawn down and the glaciers grow even more. Model studies show that the orbital changes alone cannot account for the size of the resulting glaciers. The greenhouse gas feedback seems to be an essential component in the process. The origin of the feedback must lie in the biogeochemical parts of the system. These are not yet known in detail, but the explanation is very plausible. So,

carbon dioxide and methane are not the original cause but rather a very important feedback mechanism that amplifies the response to the orbital trigger (sometimes referred to as a pace maker).

4. Earth experienced large climate changes long before human influences. Who or what caused those? *Answer:* On geological time scales, the processes of mountain building (lots of volcanism and carbon dioxide buildup) and subsequent erosion of rocks (acid rain from the carbon dioxide) on the mountainsides can modulate the carbon dioxide concentration drastically. This is undoubtedly the origin of much of the past greenhouse effects on the planet.

5. What about the Medieval Warm Period (centered around A.D. 1000), when Greenland was colonized and was possibly actually green, but abandoned in the 1300s. Wasn't that warmer than the present? *Answer:* While we cannot be certain of the existence of a Medieval Warm Period, it appears that there was a warm period in the Northern Hemisphere, but it is less likely in the Southern Hemisphere. Current studies, however, indicate that it was not warmer than the present. Greenland might have had less ice cover at that time, and it might be so again in the next century or so.

6. Climate models do not get the trend right in the upper troposphere according to satellite data. Doesn't this discredit the theory? *Answer:* This problem has been around for a few decades. Improvements in the models have helped resolve it, but more important, the early interpretation of the satellite data has been shown to underestimate the warming in the upper troposphere. The bias was extremely hard to find and took many years to unravel. Further examination of weather balloon observations has found additional errors, and it is looking like the models have been right all along.

7. How about the sun? Doesn't it change its brightness from time to time? *Answer:* The sun's irradiance does vary about 0.1 percent with the 11-year sunspot cycle, but measurements with satellites taken outside Earth's atmosphere show that it is very unlikely that a trend exists over the last 30 years, just exactly when the planetary warming has been the most rapid. We cannot say as much about earlier periods in Earth history, but some evidence suggests that the brightness has changed over millennial time scales. Because of the relatively small magnitude of its variations

and the timing of the recent warming, the sun seems to be playing at most a minor role in recent climate change.

8. Question from a spectroscopist: Since the spectral lines of carbon dioxide and the other greenhouse gases are discrete, once the concentration is large enough that the absorption lines are saturated, they no longer have an effect. How then can increasing the concentration further raise the temperature? *Answer:* Let's translate this to a real greenhouse in a typical back yard. The greenhouse glass already absorbs 100 percent of the infrared. So what if we double its thickness? It is correct that nothing happens to the temperature at the ground underneath the greenhouse. Getting back to the atmospheric problem, some of the sharp spectral lines do saturate, but when carbon dioxide is doubled, the infrared radiation to space is now emitted from a higher level in the atmosphere where the temperature is lower. This means that its total emission to space is reduced. Since the amount of incoming solar radiation is unaltered by the doubling, the planet's temperature now must increase about 1°C to maintain balance.

THE ASSESSMENT PROCESS

As the science of climate change progresses, with thousands of scientists grinding away, the state of the art must be periodically assessed, both from the engineering perspective (to address societal needs) and from the scientific perspective, in which the investigator seeks answers to deeper questions. Other issues arise, such as anomalies like the disagreement between satellite data and the modeling in Question 6. These need to be delineated and strategies developed to reconcile them within the climate science paradigm. So far this approach has been successful, as there are no serious outstanding anomalies at this time. These periodic inspections are called climate assessments. They come in a variety of forms. For example, the National Academy of Sciences frequently assembles committees to look into scientific questions and assess the status of various fields of medicine, engineering, and even climate studies. Numerous committees have taken on portions of this task over the last three decades. Besides these rigorous assessments, the professional societies with which climate scientists affiliate themselves often issue statements on matters of public interest that reflect the thinking inside the community of scholars in that particular field. For example, the American Geophysical Union (about 35,000 members) and the American Meteorological Society (about 11,000 members) have issued very strong statements.

In the case of climate science, we look not only to the assessments mentioned above but also to the Intergovernmental Panel on Climate Change (IPCC), under the auspices of the United Nations. It was charged with providing periodic assessments of climate science with special attention to possible climate change and its impacts on society. The First Assessment Report of the IPCC was published in 1990, followed by another every five or six years. The latest assessment report was published in 2007 (IPCC 2007). Much of the rest of this chapter is a summary of that report.

It is important to understand the IPCC process. Each report consists of the findings of three working groups: Working Group 1 assesses the current status of the science, Working Group 2 considers impacts of climate change given the findings of Working Group 1, and Working Group 3 examines steps that might be taken to mitigate climate change. Each working group report consists of 10 or so chapters of approximately 100 double-column pages. Each of the chapters has several lead authors and 20 or so coauthors. In assembling information for the chapters, numerous workshops and open forums are held. As the report is cast into draft form, it is sent out to anonymous referees for comment. Also, the draft report is made available on the Web. The 2007 report received some 30,000 comments, and every one was addressed in some form. Ultimately, the report summaries are read and reviewed by the political representatives of the 120 countries involved, followed by unanimous approval.

IPCC FINDINGS

Working Group 1, in Assessment Report 4, reviews all aspects of the status of the science of climate change research, with chapters on past climates, natural variability, current observations, climate models, and future climate projections. Selected conclusions from the report may be summarized as follows:

1. The concentration of carbon dioxide has been increasing and is now higher than it has been in many thousands of years. The bulk of the increase over the last century can be attributed to the burning of fossil fuels, certain industrial processes, and land use changes.
2. Understanding of climate change has improved over the last five years. It can now be stated with a very high level of confidence that human activity has been a primary influence in the warming (Fig. 1.3).

3. Warming of the earth is unequivocal from a variety of observations (Fig. 1.4).

4. Research on past climates suggests that the present warmth exceeds that of the last thousand years, and the last time it was warmer was during the previous interglacial, 125,000 years ago, when sea level was 4–6 meters (13–20 feet) higher.

5. In a stronger statement than that of the previous report, the fourth assessment states that the warming is "very likely" to be due to human activities.

6. For the next few decades we can expect a warming of about 0.2°C (0.4°F) per decade.

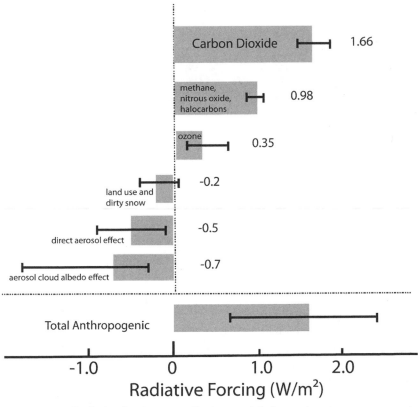

FIGURE 1.3. Radiative forcings contributing to global warming since A.D. 1750, in watts per meter squared. The I-bars represent uncertainty ranges in the estimates. Adapted from the Fourth Assessment Report of the Intergovernmental Panel on Climate Change (IPCC 2007).

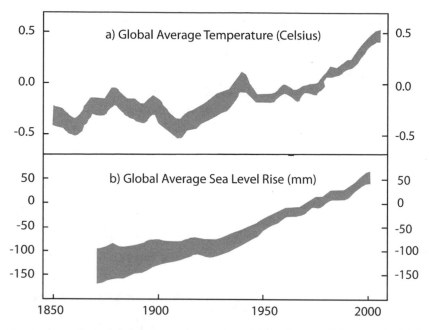

FIGURE 1.4. Past global average temperatures (a), in degrees Celsius, and global average sea level (b), in millimeters. Values are departures from the 1961–2000 average. The vertical width of the gray swatches represents the range of uncertainty. Adapted from the Fourth Assessment Report of the Intergovernmental Panel on Climate Change (IPCC 2007).

Here we focus on the projections. Such projections necessarily depend on the future record of emissions of greenhouse gases and other changes in climate drivers (e.g., land use). The IPCC has developed a suite of alternative future societal behaviors for input into climate simulation models, and some of these are briefly described in the report's *Summary for Policymakers*. In the results from simulations under various scenarios, the A2 scenario leads to the most warming (Fig. 1.5). It is a future with "very rapid economic growth, global population that peaks in mid-century and declines thereafter, and the rapid introduction of new and more efficient technologies." Even if carbon dioxide levels are artificially fixed at A.D. 2000 levels, there is still some warming due to the thermal inertia of the system (not shown).

Simulations by about 20 climate models have been used to average the change in precipitation over the decade 2090–2099 and compare it with the 1980–1999 average (Fig. 1.6). Note that the subtropics in both hemispheres show a decrease in precipitation. A number of papers, based on both models and data from the recent decades, now indicate that the storm tracks in

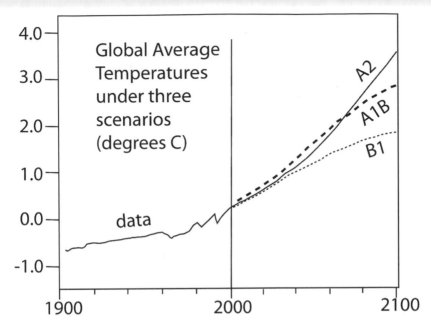

FIGURE 1.5. Projection of global average temperature for three scenarios of green-house gas emissions over the past and current centuries. A2 (solid curve) is a near business-as-usual scenario. A1B (dashed curve) and B1 (dotted curve) employ smaller emissions. A brief definition of the scenarios is given in the 2007 IPCC Summary for Policymakers. The range of uncertainty for the climate simulations is about plus or minus 0.5°C; the uncertainty in the emission scenarios is another plus or minus 1°C. Adapted and simplified from the Fourth Assessment Report of the Intergovernmental Panel on Climate Change (IPCC 2007).

winter will recede slightly toward the poles in each hemisphere under forced climate change (data-based: Archer and Caldeira 2008; Seidel and Randel 2007; model-based: Yin 2005; Kushner et al. 2001; Bengtsson et al. 2006). In particular, one study (Seager et al. 2007) focuses on the Southwestern United States (land areas from 95 degrees W, at about Interstate Highway 35, to the West Coast and from South Texas to Kansas), making use of the same simulation models as Assessment Report 4 to examine the changes in precipitation and evaporation over the region. It finds that the region will become drier toward the end of the century, with the normal climate comparable to the drought of the 1950s (Fig. 1.7). This suggests that use of the drought of the 1950s (the so-called drought of record) may not be appropriate for future water resource planning, since the study implies that large swings above and below the current drought of record will be common.

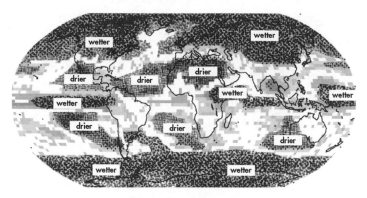

FIGURE 1.6. Simulations of December-January-February precipitation changes for the end of the century, compared with the present. Results were used from about 20 climate models. June-July-August conditions are similar. Note the band of drier (cross-hatched) conditions in the subtropics in both hemispheres, part of which overlaps West Texas. White areas indicate that less than 66 percent of the models agree on the direction of the change. Adapted from the Fourth Assessment Report of the Intergovernmental Panel on Climate Change (IPCC 2007).

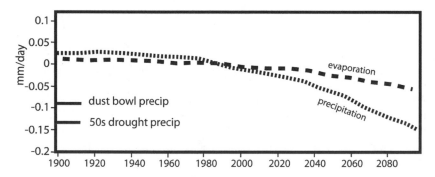

FIGURE 1.7. Precipitation and evaporation changes in the southwestern United States, 1900–2080. Simulations from 21 models (IPCC 2007) were averaged over all land between 125 and 95 degrees W and between 25 and 40 degrees N. The curves are smoothed estimates adapted from the original paper by Seager et al. (2007). Large fluctuations of 10-year drought and wet cycles are to be superimposed on these mean curves. Schematic adapted from Seager et al. (2007).

Much interannual variability of precipitation and temperature in the greater Texas region is accounted for by persistent sea surface temperature distributions in the Pacific. For example, it has been well known for many years that the El Niño and La Niña cycle (often referred to as ENSO, for El Niño–Southern Oscillation) in the Pacific modulates winter rain in Texas.

Winters are usually wet and cool during El Niños and dry and mild during La Niñas. (Excellent tutorials on ENSO can be found at www.cpc.noaa .gov and www.pmel.noaa.gov.) There is evidence that other sea surface temperature changes in the Pacific can lead to curious changes in the storm tracks crossing the U.S. mainland. For example, model studies (Schubert et al. 2004) indicate that sea surface temperature anomalies played a role in the droughts of the Dust Bowl in the 1930s. Persistent anomalies in the Pacific sea surface temperatures are part of the natural variability of the climate. Anomalies such as ENSO recur every 2–5 years, but no two are exactly alike. Other anomalies, such as those linked to the decade-long droughts of the last century, come and go with less regularity. Some might be attributable to sea surface anomalies, but others might originate with a confluence of otherwise benign circumstances leading to a perfect-storm situation.

Texas and the southeastern part of the United States tended to cool over much of the last century, until the last few decades (Chapter 2), when even this region began to fall into line with the global trend. Climate model evidence suggests a linkage of this cooling to the increasing sea surface temperatures in the tropical Pacific (Robinson et al. 2002). The modeling studies cited here do not use actively coupled ocean models; rather, they specify sea surface temperatures in a variety of numerical experiments to unravel cause-and-effect relationships. During El Niños, Atlantic hurricanes tend to be suppressed, further influencing the climate of Texas.

The present mean annual distribution of tropical precipitation (diagrammed in Fig. 1.8) features the Hadley Cell, with rising air at the equator and sinking air at about 30 degrees N and S, resulting in heavy rain along or near the equator and dry conditions in the subtropics. Note that the major world deserts occur in these zones (white areas in Fig. 1.8).

The west-east gradient in rainfall across the Texas region is apparent in the seasonal march of precipitation across the continental United States and tropical areas nearby (Fig. 1.9). There is heavy rain in the east and dry conditions in the west. This is undoubtedly due to the availability of moist air from the Gulf of Mexico in the east and the lack of moisture in air descending from the west. Other familiar features stand out in the seasonal data: the heavy winter rain in the Pacific Northwest, the Southwest monsoon coming up the western coast of Mexico in summer, the intensification and northward migration of the equatorial rains in the Northern Hemisphere in the summer. The rainfall in Texas shows a sharp west-to-east gradient in January through May, but the pattern breaks down during summer as localized convective storms tend to dominate the precipitation throughout the High Plains. What happens to this picture as the globe warms?

FIGURE 1.8. The mean annual distribution of present precipitation (right panel) from the Tropical Rainfall Measuring Mission. Air rises along the equator, giving rise to heavy precipitation. It tends to sink in the latitude bands around 20 to 30 degrees N, where dry descending air contributes to most of the world's deserts (a circulation known as the Hadley Cell). The map of satellite data indicates a band of heavy precipitation encircling the globe near the equator, with very dry conditions (white areas) in the subtropics. Data available from the TRMM website, *http://trmm.gsfc .nasa.gov.*

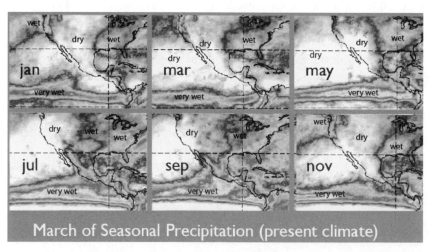

FIGURE 1.9. The seasonal cycle of rainfall for North and Central America. Note the sharp west-east gradient of rain rate in the Texas region and the way it changes over the course of the year.

IMPLICATIONS FOR THE UNITED STATES

Virtually all of the projections from the various scenarios indicate warming for the continental United States, much of it slightly more than the global average. In the western and central portions of the country, the projected changes are centered on an increase of 4.0°C (7.2°F), from 1950 to the end of the century, with model differences of about plus or minus 1.0°C, or 1.8°F (Fig. 1.10). Similar changes are projected for the eastern United States, with even larger changes in temperature expected for most of Canada (not shown here). From this we can conclude that the continental United States will be about 4.0°C warmer than it was in 1950.

The report estimates that over the next century, northern portions of the continent will experience more rain and southern portions will receive less (Fig. 1.11). This pattern is consistent with the idea of the storm tracks receding toward the North Pole and the accompanying sinking branch of the Hadley Cell also expanding toward the north. The estimates of summer precipitation suggest that the sinking branch of the Hadley Cell now extends roughly to the Canadian border, with diminished rain throughout the continental United States.

ENERGY, WATER, AND CLIMATE CHANGE

According to simulations from Assessment Report 4, mean annual temperatures for Texas are projected to rise about 4.0°C (7.2°F), plus or minus 1.0°C (1.8°F), from the last two decades of the twentieth century to the last two of the twenty-first (Fig. 1.10). Most of Texas is likely to experience

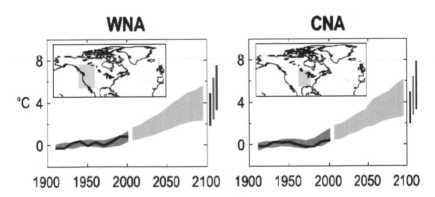

FIGURE 1.10. Warming of the continental United States, centered on an increase of 4.0°C, from 1950 to 2100. Results are from simulations of about 20 climate models for Western North America (WNA) and Central North America (CNA). Regions are indicated by the gray boxes in the maps. The solid black curve is the data, and the dark gray swath covering it is the range of simulations (90 percent fall within). The light gray swath (after the year 2000) indicates the same range of the projections. The vertical lines on the right indicate the range for each of three scenarios. Adapted from Fig. 11.11 of Working Group 1, in the Fourth Assessment Report of the Intergovernmental Panel on Climate Change (IPCC 2007).

AR4 Precipitation Estimates for the next Century

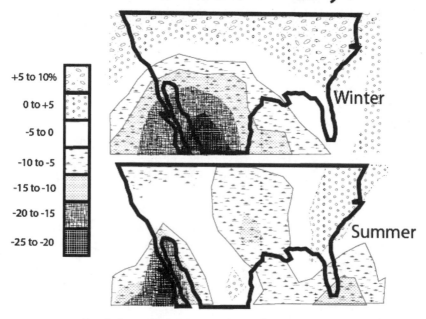

FIGURE 1.11. Simulations of precipitation change from the last two decades of the twentieth century to the last two of the twenty-first. The upper panel depicts the percentage of change in mean winter (December-January-February) precipitation; the lower panel is the change in summer precipitation. Adapted from Fig. 11.12 of Working Group 1, in the Fourth Assessment Report of the Intergovernmental Panel on Climate Change (IPCC 2007).

much higher air-conditioning bills. Since heating bills are much less than cooling bills, the savings in winter are not likely to balance the increased expense in summer.

Estimating future precipitation is more problematic. The global climate model results suggest that virtually all of Texas will get less precipitation in both winter and summer (Fig. 1.11). The decrease might be as little as 5 percent or as much as 15 percent, especially in West Texas in winter. According to the model projections, the rain in summer might not be diminished as much, thanks to the Southwest monsoon as the convection creeps up the west coast of Mexico in summer (Fig. 1.9). There is large variability, however, among the model projections for both winter and summer (as discussed in the following chapter).

The next chapter also notes that over the past century, precipitation increased in East Texas but was unchanged in West Texas. Why, then, do models predict drying? Are they missing such important fine-scale features as low-level jets that efficiently convey moisture northward from the Gulf of Mexico? This moisture in turn feeds the storms so common to the East Texas area. The mechanisms for precipitation are complex in this interesting region, and the lack of spatial resolution in our current models might mislead.

Obviously, water resources, energy and food prices, population increases, threats to ecological habitats, and climate change interact in complex ways. If Texas gets less precipitation and higher temperatures, we can expect more evaporation and less runoff into rivers and less recharge of aquifers. For example, if Texas is drier, with prolonged droughts on top of the drier normals, we can expect interruptions by power plants using stream-derived water as a coolant. West Texas may find itself shifting from an agricultural economy to a wind energy economy because of an inadequate supply of low-cost water.

CONCLUSIONS AND FUTURE EXPECTATIONS

Climate science really emerged as a thriving enterprise in the 1970s with the sudden increase in scientific attention to the problem. The new awareness was rooted in the emerging prospects for attacking and solving many interesting problems in past and future climates both on Earth and on other planets. This stimulated the collection of global data sets, along with advances in quality control and archiving. The speed and capacity of large digital computers and the adaptation of weather forecasting models to the climate problem enabled scientists to attack new problems in climate research. Other technological advances, such as instrumented research aircraft and ships, led to exciting new field programs to check and improve models. Scientists were attracted from neighboring fields, blending their expertise with that of meteorologists and oceanographers. The result has been an explosion of research on past and future climates. The climate science paradigm is being pursued at a fever pitch, and the state of the art is regularly assessed by many scientific organizations; the Intergovernmental Panel on Climate Change provides comprehensive assessments every five years or so.

Many conclusions in climate science now appear to be robust. These include the general warming of the planet (1.5–4.5°C, or 3–8°F, during the coming century), driven mainly by increasing concentrations of greenhouse gases. There is and will continue to be more warming at the poles than elsewhere, and the global hydrological cycle is very likely to intensify between a few percent and 7 percent per degree Celsius of warming. If no catastrophic melting occurs on Greenland or Antarctica, we can expect sea level to rise 1–2 feet. All the ice on Greenland represents a rise in sea level of about 18 feet, and the same for the West Antarctic ice sheet. They are unlikely to melt during this century, but there is a small probability of it. Because of the potential consequences, it cannot be taken lightly.

The forecasting of future climate details for a region as small as Texas is still problematic. This is even more difficult for Texas because of its unique location, bordering to its south the contrasting surfaces of arid Mexico and the warm waters of the Gulf of Mexico. These conditions combine with the prevailing winds across Texas, coming from the south most months of the year. The state is also located at a latitude such that winter weather (cold and warm fronts and rainy passages) crosses most months of the year; these so-called storm tracks are subject to some change as global warming proceeds. Storm tracks are likely to recede northward with global warming, making the passage of fronts across Texas less frequent in spring and fall. Texas is also situated next to New Mexico and Arizona, both of which experience a summer monsoon creeping up the west coast of Mexico. West Texas could benefit from the summer monsoon, if it strengthens. Texas winters are also strongly influenced by the cycle of El Niño and La Niña. Typically El Niño brings wet winters to Texas. Hurricanes and tropical storms impinge on Texas coasts. The frequency and intensity of both El Niños and tropical storms might change with global warming. The occurrences and interactions of all these factors make for a precarious forecast for precipitation. There can be legitimate differences of opinion; this author opts for more rain in the eastern part of the state and less in the west, but strictly speaking, the jury is still out.

What about downscaling? This is the process of taking the projection of a global model on a coarse grid (e.g., Fig. 1.2) and using either a statistical regression scheme to extrapolate to smaller scales or actually embedding a finer-scale model within the coarser grid. In either case, we run into the problem of choosing a coarse-scale model to use for the input. If we feed poor coarse-scale information to a fine-scale model, there may be very little information added; but if topography or some other factors are important, there could be genuine value in the downscale. The efficacy of downscaling is still an open question in climate modeling and the focus of current research.

The most formidable problems facing climate research are connected with convection, because important features of these bursts of rising air are usually far smaller than our grid lattices (e.g., localized tropical precipitation phenomena). Great efforts are being expended along these lines. The range of uncertainty is due mainly to our lack of quantitative understanding of the feedbacks affecting the problem. It may be many years before these are cleared up, but I expect serious progress on the frequency and intensity of tropical storms and El Niño in the next decade. Other consequences of warming will also be better understood, including impacts on the ice sheets of Greenland and Antarctica and the hydrological cycle. There is a good chance that we can clarify the situation on future precipitation in Texas in the next decade as increased computer power allows model grids to become denser.

REFERENCES

Archer, C. L., and K. Caldeira, 2008. Historical Trends in the Jet Streams. *Geophysical Research Letters* 35. DOI: 10.1029/2008GL033614.

Bengtsson, L., K. Hodges, and E. Roeckner, 2006. Storm Tracks and Climate Change. *Journal of Climate* 19:3519–3543.

Budyko, M. I., 1968. The Effect of Solar Radiation Variations on the Climate of the Earth. *Tellus* 21:611–619.

Dessler, A. E., and E. A. Parson, 2010. *The Science and Politics of Global Climate Change.* 2d ed. Cambridge University Press, New York.

IPCC, 2007. Summary for Policymakers. In: *Climate Change 2007: The Physical Science Basis.* Contribution of Working Group 1 to the Fourth Assessment Report of the Intergovernmental Panel on Climate Change. Cambridge University Press, Cambridge, U.K., and New York.

IPCC, 2007. Summary for Policymakers. In: *Climate Change 2007: Impacts, Adaptation, and Vulnerability.* Contribution of Working Group 2 to the Fourth Assessment Report of the Intergovernmental Panel on Climate Change. Cambridge University Press, Cambridge, U.K., and New York.

IPCC, 2007. Summary for Policymakers. In: *Climate Change 2007: Mitigation of Climate Change.* Contribution of Working Group 3 to the Fourth Assessment Report of the Intergovernmental Panel on Climate Change. Cambridge University Press, Cambridge, U.K., and New York.

Kuhn, Thomas, 1970. *The Structure of Scientific Revolutions,* 2d ed. University of Chicago Press.

Kushner, P. J., I. M. Held, and T. L. Delworth, 2001. Southern Hemisphere Atmospheric Circulation Response to Global Warming. *Journal of Climate* 14:2238–2249.

Plass, Gilbert, 1956. The Carbon Dioxide Theory of Climate Change. *Tellus* 8:868–875.

Robinson, W. A., R. Reudy, and J. E. Hansen, 2002. General Circulation Model Simulations of Recent Cooling of the East-central United States. *Journal of Geophysical Research* 107. DOI: 10.1029/2001JD001577.

Schubert, S. D., M. J. Suarez, P. J. Pegion, R. D. Koster, and J. T. Backmeister, 2004. Causes of Long-Term Drought in the U.S. Great Plains. *Journal of Climate* 17:485–503.

Seager, R., M. Ting, I. M. Held, Y. Kushnir, J. Lu, G. Vecchi, H.-P. Huang, N. Harnik, A. Leetmaa, N.-C. Lau, C. Li, J. Velez, and N. Naik, 2007. Model Projections of an Imminent Transition to a More Arid Climate in Southwestern North America. *Science* 316:1181–1184.

Seidel, D. J., and W. J. Randel, 2007. Recent Widening of the Tropical Belt: Evidence from Tropopause Observations. *Journal of Geophysical Research* 112. DOI: 10.1029/2007JD008861.

Sellers, W. D., 1969. A Climate Model Based on the Energy Balance of the Earth-atmosphere System. *Journal of Applied Meteorology* 8:392–400.

Worrall, J., 2003. Normal Science and Dogmatism, Paradigms and Progress: Kuhn versus Popper and Lakatos. P. 298 in: *Thomas Kuhn*. T. Nickles (ed.). Cambridge University Press, Cambridge, U.K.

Yin, J., 2005. A Consistent Poleward Shift of the Storm Tracks in 21st Century Climate. *Geophysical Research Letters* 32. DOI: 10.1029/2005GL023684.

CHAPTER 2

The Changing Climate of Texas
John W. Nielsen-Gammon

Texas is the second-largest state in the United States, with a total land area of 261,914 square miles. Its size and geographical location combine to give it a diverse climate, with a wide variety of local and regional climatic influences. Before discussing the historical record of past climate change in Texas and possible future changes as a result of global warming and other factors, this chapter will give an overview of Texas climate and how the climate is determined by topographic and other characteristics.

OVERVIEW OF TEXAS CLIMATE

The Texas climate is strongly influenced by three large geographical features. The first is the Rocky Mountains, or more broadly, the North American Cordillera. This set of mountain ranges and plateaus, extending from Alaska through western Canada and the United States to Mexico and Central America, presents a formidable barrier to air traveling from west to east or from east to west. Except in the El Paso area within the cordillera, winds from the north and south are much more common in Texas than winds from the east or west.

The other two substantial geographical features are the central and eastern North American continent and the Gulf of Mexico. Both form a barrier-free pathway for the movement of air into Texas, but the characteristics of the air depend on the route taken. The Great Plains form an unbroken, relatively flat landmass stretching from the Arctic Circle to Texas. Cold air masses that appear on the eastern side of the North American Cordillera are channeled southward, traversing the distance between the U.S.-Canadian border and Texas in about a day and a half. The Gulf of Mexico, on the other hand, serves as a moisture source for Texas. Much of the moisture crossing the coastline originates from the Caribbean Sea or the open Atlantic, but the air picks up additional moisture from the Gulf of Mexico before reaching Texas. The Gulf of Mexico also serves as a moderating influence on temperature, since its surface temperatures do not change nearly as much from day to day or from winter to summer as land surface temperatures.

The basic climatic patterns in Texas are fairly simple: annual mean temperature increases from north to south, and annual mean precipitation increases from west to east. Typical daily minimum temperatures in January range from about 20°F in the northern Panhandle to about 50°F near the mouth of the Rio Grande (Fig. 2.1). The north-south contrast is completely absent for maximum temperatures in July (Fig. 2.2). Hotspots are found along the Rio Grande and Red River, while the coolest summertime temperatures are found in the mountains of West Texas.

The broad east-west variation in precipitation is caused by the Gulf of Mexico, which can more readily supply moisture to the eastern half of the state than to the western half (Fig. 2.3). Exceptions to this pattern include the Balcones Escarpment, on the edge of the Edwards Plateau west of San Antonio, where precipitation is enhanced by air being forced upward by the topography as it travels northward, and West Texas, where precipitation is more common in the higher terrain than the lower because of upslope flow and summertime thunderstorms.

Regional and Seasonal Climatic Variations

For the purposes of this chapter, Texas may be subdivided into eight climatic regions (Fig. 2.4), each with its own distinct seasonal weather patterns. For each region, seasonal variations in precipitation are depicted at one key long-term station (with roughly 100 years of data) using box-and-whisker charts. These charts show the median monthly precipitation (thick horizontal line), the values between the lowest quarter of all months and the highest quarter of all months (box), and the values between the lowest 5 percent and the highest 5 percent of all months (vertical lines). For example, in El Paso in June (Fig. 2.5), a third of an inch of precipitation or more occurs about half the time (the median line), zero precipitation occurs almost a quarter of the time (the bottom of the box), and more than 5 percent of all Junes produce greater than 2 inches of precipitation (the top of the vertical line).

Because of the skewed nature of precipitation, median precipitation is almost always smaller than average or normal precipitation. In other words, precipitation is more likely to be below normal than above normal.

Far West Texas has a well-defined wet season and dry season (Fig. 2.5). The dry season is November through May, and the wet season is June through October. The peak rainfall months of July and August are a consequence of the Southwest monsoon, a flow pattern that brings moist tropical air and convection to northwestern Mexico, Arizona, New Mexico, and the western edge of Texas. Long-term changes in rainfall in Far West Texas are closely related to changes in the pattern of the Southwest monsoon. During the dry season, entire months without precipitation are not uncommon.

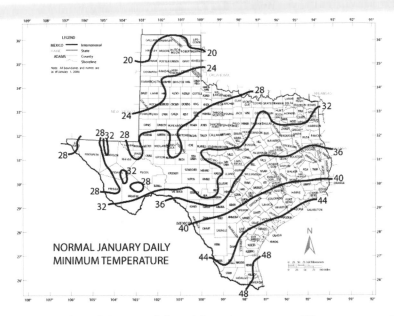

FIGURE 2.1. Normal January daily minimum temperature (°F), 1971–2000. Redrawn after graphics created February 20, 2004, by the PRISM Group, Oregon State University, *http://prismclimate.org*.

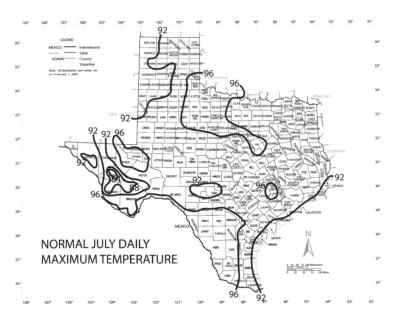

FIGURE 2.2. Normal July daily maximum temperature (°F), 1971–2000. Redrawn after graphics created February 20, 2004, by the PRISM Group, Oregon State University, *http://prismclimate.org*.

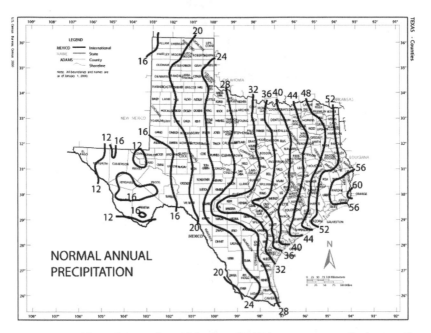

FIGURE 2.3. Normal annual precipitation (inches), 1971–2000. Redrawn after graphics created February 20, 2004, and June 16, 2006, by the PRISM Group, Oregon State University, *http://prismclimate.org.*

The Panhandle and Plains also receive a share of the July and August Southwest monsoon precipitation, but even more precipitation falls in May and June (Fig. 2.6). This is the period of time when the West Texas dryline is active. The dryline marks the boundary between humid, potentially unstable air from the Gulf of Mexico and hot, dry, descending air from the interior Southwest. Severe thunderstorms are favored to form along and to the east of the dryline, sometimes producing tornadoes and large hail. The monsoon precipitation is more regular than the dryline precipitation, as seen by the relatively short range of precipitation for August compared with earlier months.

In West Central Texas, the springtime precipitation maximum occurs squarely in May (Fig. 2.7), a bit earlier than in the Panhandle and Plains. Without moisture from the Southwest monsoon, July is one of the drier months of the year in this part of Texas. This dryness, in terms of precipitation, is exacerbated by the higher temperature during July, making July the driest month of the year in terms of precipitation minus evaporation. Wintertime precipitation, although still much lower than in late spring or

FIGURE 2.4. Climatic regions of Texas. Stars indicate representative long-term climate observing stations within each region: El Paso in Far West, Plainview in Panhandle and Plains, Ballinger in West Central, Weatherford in North Central, Boerne in South Central, Rio Grande City in South, Danevang in Southeast, Marshall in East.

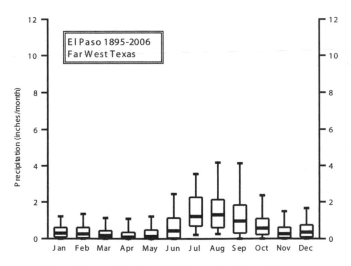

FIGURE 2.5. Monthly precipitation in El Paso, Far West Texas, 1895–2006. Horizontal line within box indicates median precipitation for that month. Bottom and top of box indicate 25th and 75th percentiles of precipitation; bottom and top of whiskers (vertical lines) represent 5th and 95th percentiles.

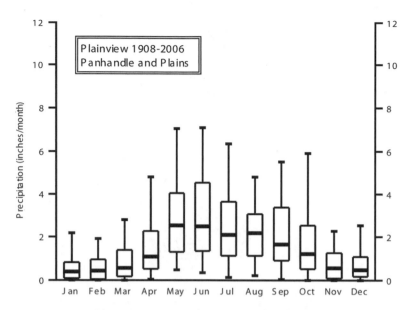

FIGURE 2.6. Monthly precipitation in Plainview, Panhandle and Plains, 1908–2006. (Box-and-whisker plot as described in Fig. 2.5.)

early fall, is much higher than in Far West Texas or in the Panhandle and Plains.

The seasonality in North Central Texas is similar to that in West Central Texas. Peaks in median precipitation occur in May and September (Fig. 2.8). Wintertime precipitation continues to increase as locations farther to the east are considered, and the likelihood of a month without rainfall, or without significant rainfall, is lower than at any other station examined thus far.

In South Central Texas, the precipitation peak in early fall is nearly as prominent as that in late spring (Fig. 2.9). The fall peak increases in prominence closer to the Gulf of Mexico, because tropical disturbances are an important source of moisture and instability.

South Texas has elements of many other climatic regions in Texas (Fig. 2.10). In common with Far West Texas, it receives very little precipitation during the wintertime. In common with North Central and South Central Texas, it has two precipitation maxima during the year, one in May or June and one in September. The September maximum is especially prominent. These climate variations are due to a combination of geographical location and wind variations. During the wintertime, South Texas lies in the rain shadow of the high terrain of Mexico as disturbances move from west to east. During the time of tropical disturbances, particularly June and

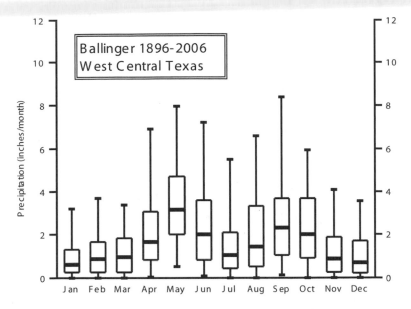

FIGURE 2.7. Monthly precipitation in Ballinger, West Central Texas, 1896–2006. (Box-and-whisker plot as described in Fig. 2.5.)

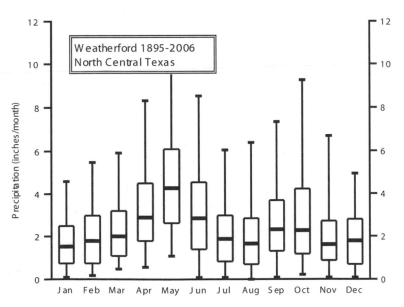

FIGURE 2.8. Monthly precipitation in Weatherford, North Central Texas, 1895–2006. (Box-and-whisker plot as described in Fig. 2.5.)

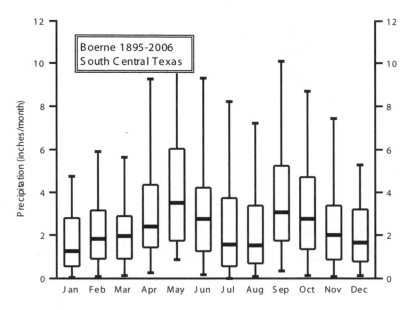

FIGURE 2.9. Monthly precipitation in Boerne, South Central Texas, 1895–2006. (Box-and-whisker plot as described in Fig. 2.5.)

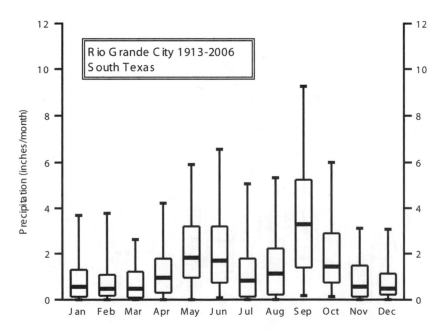

FIGURE 2.10. Monthly precipitation in Rio Grande City, South Texas, 1913–2006. (Box-and-whisker plot as described in Fig. 2.5.)

September, disturbances tend to move from southeast to northwest, and the easterly winds ahead of storms favor upslope flow and precipitation.

The annual pattern of precipitation in Southeast Texas is similar to that in South Texas, except that about 2 inches more falls on average in each month in Southeast Texas (Fig. 2.11). The relatively large amount of precipitation in Southeast Texas means that there is enough moisture to keep the ground damp during most of the year, with significant water stress typical only during July and August.

In East Texas (Fig. 2.12), the precipitation pattern is almost the exact opposite of that in Far West Texas. In East Texas, the climatologically wettest month of the year is December, and the driest months are July, August, and September. East Texas is far enough north to be affected by frequent traveling wintertime disturbances, and far enough east that there is generally ample moisture available when the disturbances arrive. Rainfall is least when temperatures are warmest, and in August, median precipitation in Marshall is actually lower than median precipitation in Plainview, in the Panhandle and Plains region.

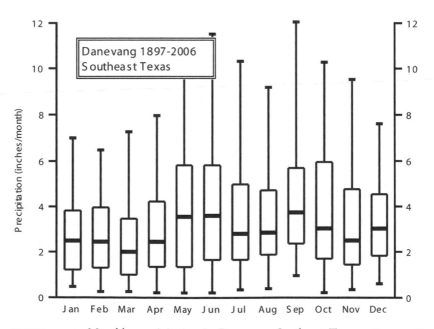

FIGURE 2.11. Monthly precipitation in Danevang, Southeast Texas, 1897–2006. (Box-and-whisker plot as described in Fig. 2.5.)

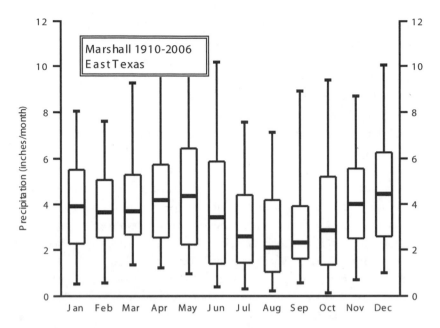

FIGURE 2.12. Monthly precipitation in Marshall, East Texas, 1910–2006. (Box-and-whisker plot as described in Fig. 2.5.)

Severe and High-Impact Weather

Texas climate, including the impact of changes in Texas climate, is defined not just by changes in average conditions but also by changes in weather extremes. One could argue that the state of Texas is the worldwide leader in the combined frequency and variety of severe and high-impact weather.

Severe weather includes tornadoes, hail, and damaging thunderstorm winds. The prevalence of all three types of severe weather generally increases in Texas from south to north and from west to east (Fig. 2.13). In South Texas and Far West Texas, significant tornadoes (those producing F2 damage or greater) come within 25 miles of any given point less than once every two decades. This does not mean that significant tornadoes are impossible there; one of the deadliest tornadoes in Texas history struck Goliad, in South Texas.

Severe weather is most common along the Red River from Wichita Falls to the Arkansas border. On the whole, hail at least three quarters of an inch in diameter and thunderstorm winds of at least 50 knots are equally common, with damaging winds slightly more prevalent in eastern Texas and hail slightly more prevalent in western Texas.

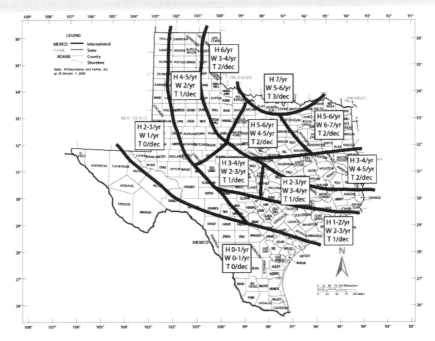

FIGURE 2.13. Climatological frequency of severe weather in Texas. Data boxes within each region indicate H: expected frequency of hail of 3/4 inch or larger diameter (events per year); W: strong thunderstorm winds of 50 knots or greater (events per year); and T: significant tornadoes of F2 or greater (events per decade). An event is defined as an occurrence within 25 miles of a given location. Spatial analyses, conducted by the National Severe Storms Laboratory (2003), are based on the period 1980–1999 for hail and wind and 1921–1995 for tornadoes.

All types of severe weather are most common in the springtime. The peak likelihood of hail is in late April in East Texas; early May in North Central, Southeast, South Central, and South Texas; late May in West Central and Far West Texas; and early June in the Panhandle and Plains. The peak likelihood for significant tornadoes is slightly earlier in the year; for thunderstorm winds it is slightly later in the year.

Hurricanes are most common in late August and September, although significant storms have struck Texas as early as June. Overall, the frequency of a hurricane passing within 75 miles of any given coastal location averages once every 7–15 years, according to the National Hurricane Center. Major hurricanes (category 3 or stronger on the Saffir-Simpson scale) pass within 75 miles of any given coastal location about once every 25–46 years.

Wind and tornadoes certainly can be major hazards, but storm surge is responsible for the greatest damage and loss of life from hurricanes. The magnitude of a storm surge depends on the strength, longevity, and size of the hurricane, as well as the configuration of the coastline.

Hurricanes can also produce copious amounts of rainfall. The intensity of the hurricane itself matters little when it comes to rainfall, and tropical storms are just as capable of producing major flooding as full-blown hurricanes. Of the 21 rainfall events in Texas between 1948 and 2002 that produced at least 20 inches of precipitation, 4 were associated with hurricanes, 5 were associated with tropical storms, and the remainder were nontropical in origin (Nielsen-Gammon et al. 2005). A memorable recent example of rainfall produced by a tropical storm was Allison in 2001, which flooded downtown Houston and caused more damage than any other tropical storm in U.S. history. Another tropical storm, Claudette in 1979, produced the greatest 24-hour precipitation total in U.S. history (43 inches in Alvin, Texas).

Other heavy rainfall events occur when deep, moisture-laden air from the south is forced upward over Texas by a slowly moving, upper-level disturbance. The unstable air produces strong convection, and the wind patterns are such that thunderstorms form repeatedly over a small area, such as north of a stationary front. The topography of the Balcones Escarpment provides additional lift for such air masses, favoring the formation of precipitation and leading to enhanced rainfall there (Fig. 2.3). In addition, the rugged, rocky soil of the Hill Country inhibits infiltration of water into the soil, producing rapid runoff and contributing to the flash-flood potential of the area.

Droughts are also a common high-impact weather event. The drought of the 1950s, the most severe long-term drought in Texas, is usually treated as the drought of record for water planning purposes. More recently, significant droughts have occurred in 1996, 1998, 2000, 2005–2006, and 2008–2009.

Because potential evapotranspiration far exceeds precipitation in the summertime throughout the state, and in all seasons in the western part of the state, most of Texas is vulnerable to short-term variations in rainfall. One or two months without precipitation in the wintertime generally has little impact, but the same rainfall deficit in the summer can be catastrophic because of the high temperatures and evaporation rates. Such short-term summertime water deficits are harmful to agriculture and ranching and can tax the ability of municipal water systems to deliver sufficient water for their customers. Lengthier water deficits (lasting six months or more) can also cause significant shortages in the supply of water.

Wildfire frequency and danger are closely tied to weather. Although one might assume that the greatest wildfire danger is associated with the most severe droughts, wildfires are most dangerous when a few months of very wet weather produce heavy vegetation growth and are followed by a few months of very dry weather that cause the fuel to dry out. When the fuel is in place, dangerous wildfires are most likely to occur on days with very strong wind and low humidity. Dust storms are favored under similar weather conditions but with longer dry periods.

Winter weather is a danger to Texas in part because it is so rare. Except for the Panhandle and Plains and the mountains of Far West Texas, significant snowstorms or ice storms are unusual. As a result, they normally bring traffic to a standstill. Ice storms, in particular, cause damage to trees and power lines and can produce power outages that last for days or weeks, particularly in East Texas. The frequency of winter weather decreases as one goes southward, but snow has been recorded in all corners of the state.

THE HISTORICAL CLIMATE RECORD IN TEXAS

Climate records in Texas go back more than a century. Some individual records exist from the middle of the eighteenth century, and by 1900 climate records through most of the state were collected as part of the United States Cooperative Observer Program (COOP). A subset of COOP stations with relatively long, homogeneous records has been designated the United States Historical Climatology Network (USHCN) (Easterling et al. 1996; National Climatic Data Center 2008).

Individual stations from the USHCN Version 2 data set illustrate this chapter's discussion of regional and seasonal climate variations in Texas (Figs. 2.5–2.12). For the purposes of examining variations over longer periods, data from all the Version 2 stations within each climate region are standardized with respect to long-term means (computed from all available data for 1895–2006) and averaged together. In order to suppress random variability associated with month-to-month changes in weather, while preserving information on seasonality, data from multiple months are averaged together and presented here.

CHANGES IN TEMPERATURE
Past Temperature Changes

The most direct manifestation of global warming is a rise in surface temperatures. Although there are many potential sources of error in surface temperature measurements that are not entirely removable through adjustment procedures (Pielke et al. 2007), interannual variations of a few tenths of a degree Fahrenheit on a regional scale and century-long variations of a degree or more Fahrenheit are probably trustworthy. This is particularly true when spatial patterns of temperature change do not correspond to known patterns of urbanization or other possible influences.

Local temperature changes are significant for agriculture, ecosystems, energy use, water supply, and other aspects of the Texas economy and way of life. Global temperature changes are far from irrelevant (they affect Texas through changes in weather patterns, sea level, moisture, and a variety of indirect influences), but they are documented and explained in publications produced by the Intergovernmental Panel on Climate Change (IPCC). Because of the importance of local temperature changes and the relative inattention to them by global panels, this section focuses on temperature changes taking place within Texas itself.

Temperatures on a global scale are strongly constrained by balances in the global energy budget, so it is relatively easy to relate global temperature changes to changes in radiative forcing, such as incoming solar radiation, greenhouse gas concentrations, and aerosols. Local temperature changes, even over decades to centuries, may also be strongly influenced by changes in regional climate patterns and sea surface temperature variations, making such changes inherently more complex.

Temperature variations during different times of the year affect society and ecosystems in different ways. Wintertime temperatures (in Texas, December through February) help determine the hardiness required of plants, energy expended on heating, and the survivability of pests. Temperatures during the swing months of March–April and October–November determine the length of the growing season and planting times. Temperatures during the hotter part of the year (May–September) strongly affect evaporation rates, energy spent on cooling, crop survival, and the tolerability of the climate.

The wintertime temperatures in Texas show considerable variability, even with nine-year averages (Fig. 2.14). Temperatures increased fairly steadily during the first half of the twentieth century, reaching a peak in the early 1950s. Thereafter, temperatures dropped, attaining twentieth-century minima in the early 1960s and early 1980s, up to 4°F colder than in the early 1950s. Since then, temperatures have warmed, and the past decade and a half has produced temperatures warmer than at any time in the past century except the early 1950s.

Because annual and longer temperature averages vary on a fairly broad spatial scale, the various regional temperatures track each other closely. The time series least like the others is from East Texas, which seems to indicate a lesser tendency toward warming over the past century.

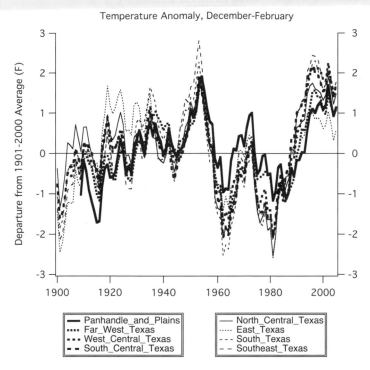

FIGURE 2.14. Departures from long-term mean for winter temperatures in eight Texas climate regions, 1900–2000. Nine-year running mean of average values of December–February mean, from U.S. Historical Climatology Network (USHCN) Version 2 data set.

In contrast, the temperatures during the swing months of March–April and October–November have much less decade-scale variability (Fig. 2.15). In other words, the length of the growing season has undergone very little change from decade to decade, until recently. The most anomalous nine-year average temperatures are those at the end of the period, which are several tenths of a degree warmer than any previous temperature maxima in all Texas climatic regions. Aside from the coinciding temperature minima in 1916, there appears to be little or no correspondence between variations in wintertime temperatures and variations in swing-season temperatures. As in winter, the East Texas temperatures are least similar to the others, again showing a tendency to be relatively warm during the early part of the century and relatively cool during the latter part of the century.

During the warm-season months of May through September, interdecadal variability is small, like the swing months, but the overall pattern is

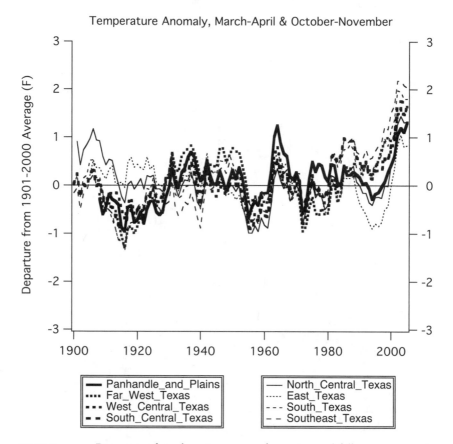

FIGURE 2.15. Departures from long-term mean for spring and fall temperatures in eight Texas climate regions, 1900–2000. Nine-year running mean of average values of March–April and October–November means, from USHCN Version 2 data set.

similar to the winter months (Fig. 2.16). Temperatures climb to a maximum in the early 1950s, then decline rather sharply. After two relative minima, temperatures increase again, and the most recent temperatures are similar to the historical maximum in the 1950s. Oddly, although maxima and minima correspond rather well between winter and summer during the first half of the twentieth century, they are almost precisely opposite during the period 1960–1990. As in the other two periods, East Texas shows much less warming over the century than is apparent in the graphs for the other climatic regions.

If the temperature observations are taken at face value, temperatures across Texas have increased fairly steadily over the past 20–30 years. This

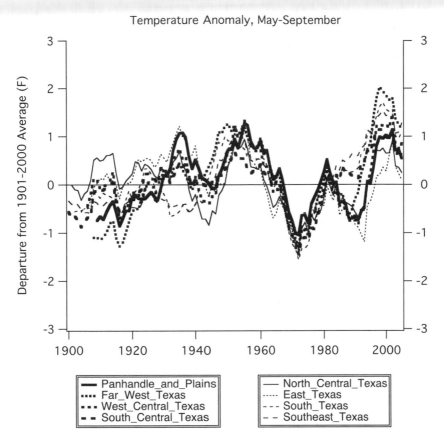

FIGURE 2.16. Departures from long-term mean for summer temperatures in eight Texas climate regions, 1900–2000. Nine-year running mean of average values of May–September mean, from USHCN Version 2 data set.

increase began, however, during a period of anomalously cold temperatures. It is only during the last 10–15 years that temperatures have become as warm as they were early in the twentieth century.

The historical temperature variations across Texas are affected much more strongly by decade-scale temperature variations than by long-term temperature trends. Nonetheless, it is useful to examine the linear trend over the past century. Despite the fundamentally different nature of decade-scale variations throughout the year in Texas, trends over the century are quite uniform across seasons. South and Southeast Texas have the largest trends, 1.1–2.0°F per century. The warming trends are more moderate from the Panhandle and Plains to Southeast Texas, 0.4–1.5°F. Only a very slight trend is present in East and North Central Texas, ranging from −0.1 to 0.8°F per century. Most of these trends are comparable to the global mean temperature trend (roughly 1.1°F per century) over the same period.

Future Temperature Changes

What does the present temperature trend imply for projections of future climate change in Texas? Knutson et al. (2006) examined the spatial patterns of temperature change throughout the globe and their reproducibility in an advanced coupled climate model. An area of negative century-long temperature trends extended from Central Texas across the southeastern United States at that time. This region was one of the few areas of cooling over the entire globe. While ensemble runs of the coupled climate model, including both anthropogenic and natural forcing, produced realistic historical simulations over most of the globe, they completely failed to reproduce the observed cooling in the south-central and southeastern United States.

Robinson et al. (2002) took a close look at this anomaly. The climate model they used was able to reproduce the anomaly, whether or not anthropogenic forcings were included, as long as the atmospheric model was driven by observed sea surface temperature patterns. Through a series of experiments, Robinson et al. (2002) found that the key region for driving the cooler temperatures in the south-central and southeastern United States was the eastern equatorial Pacific Ocean, where warm sea surface temperatures led to increased water vapor and cloud cover over the southern United States, thereby reducing temperatures throughout the year. Thus, it seems that regional temperature change patterns within Texas are probably unreliable indicators of future climate change, because of the strong influence of variations in patterns of sea surface temperatures.

The amount of future warming due to changing greenhouse gas concentrations has been estimated by the RCPM method (Regional Climate-Change Projection from Multi-Model Ensembles; Tebaldi et al. 2005; National Center for Atmospheric Research 2008), courtesy of Seth McGinnis of the National Center for Atmospheric Research. This particular estimation uses the archived set of climate simulations for the A1B emissions scenario and computes changes over a 10-grid-cell domain covering Texas (Fig. 2.17). The technique produces probability distributions of future climate, assuming the emissions scenario is accurate, by weighting the climate models according to their ability to simulate current climate and their similarity to other models in terms of long-term trends.

The RCPM data and analyses are provided by the Institute for the Study of Society and Environment at the National Center for Atmospheric

Region used in computation

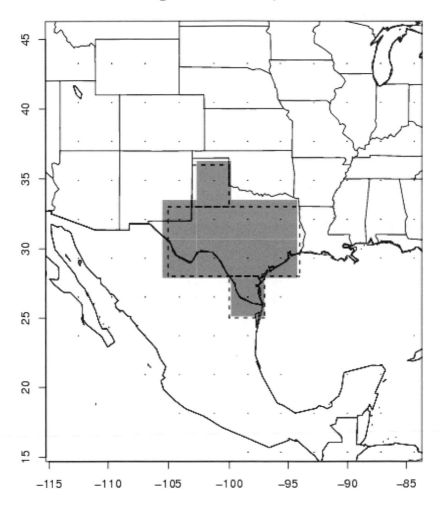

FIGURE 2.17. Averaging area used for estimation of projected temperatures and precipitation in Texas, from A1B scenario.

Research, based on model data from the World Climate Research Program's Coupled Model Intercomparison Project, phase 3, multi-model data set. More information about the RCPM analysis can be found at http://rcpm .ucar.edu.

The projected temperature trends are computed relative to a simulated 1980–1999 mean (Fig. 2.18). The projections indicate an increase of about 1°F for the 2000–2019 period, 2°F for the 2020–2039 period, and close to 4°F for the 2040–2059 period. With the first eight years of the 2000–2019 period being the warmest such period on record, the projection for that period is well on its way to reflecting reality. By 2020–2039, the chance of temperatures in Texas actually being cooler than 1980–1999 is less than 5 percent, according to the climate models, and most of the projections fall well outside the range of natural variability.

It should be kept in mind that these projections assume no massive volcanic eruptions, sudden decreases in solar output, or other such events that would have a cooling effect on temperatures. Also, the projections assume no explicit action to combat global warming through reduction of greenhouse gas emissions, but reductions of the magnitude being presently contemplated would have little effect on global temperatures over the next 30 years.

CHANGES IN PRECIPITATION
Past Precipitation Changes
Unlike temperature, precipitation is not expected to increase globally. As shown in Chapter 1, climate model simulations indicate that precipitation

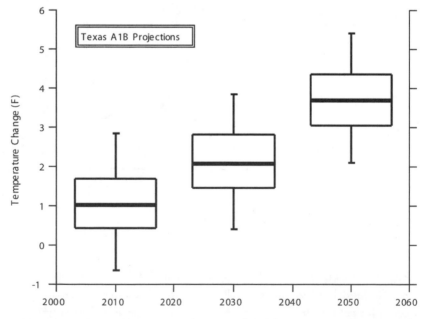

FIGURE 2.18. Statistical estimates of ranges of future temperatures in Texas relative to 1980–1999 baseline, according to Bayesian analysis of global climate models driven by the A1B forcing scenario. Each estimate is for a 20-year average. RCPM data and analyses are 2006 University Corporation for Atmospheric Research (All Rights Reserved).

should generally increase along the equator and at high latitudes and generally decrease in other parts of the tropics and subtropics. This banded pattern of precipitation changes corresponds fairly closely to the pattern of observed changes, with one exception. Around 30 degrees N, the global climate models predict that a decrease in precipitation should have occurred, but the observed change of precipitation at this latitude has been positive (Zhang et al. 2007). Texas lies within this latitude band.

For examination of the precipitation record in Texas, months are grouped according to similar long-term precipitation trends. To enable comparison of similar impacts in different regions of the state, precipitation is expressed as a fraction of the local mean for the period of record, defined as for temperature above. Nine-year averages are plotted.

Precipitation for December–March has increased substantially throughout the state (Fig. 2.19). The largest percentage increase, 45 percent per century, took place in the Panhandle and Plains, but since this is the dry season for that area, the impact on water supplies is minimal. The steadiest increase, with the least variation per decade, has been in East Texas, which receives a greater amount of precipitation during December–March than any other Texas climatic region, in both absolute and relative terms. There, the increase has been 14 percent per century.

Wet spells and dry spells tend to coincide throughout the state over most of the period of record. The primary exception is the most recent decade, when Far West and South Texas received close to their long-term average amount of precipitation, while North and East Texas received 20–50 percent more than their long-term average. Note that the use of nine-year averages masks shorter-term precipitation variations. The Panhandle and Plains, for example, alternated very wet and dry years during this period.

During the period April–July, precipitation shows little long-term trend and less interdecade variability than in December–March (Fig. 2.20). Unlike the wintertime, in these months the southern and northern parts of the state seem to vary nearly independently over most of the past century. In the most recent years, the regions receiving an increase in precipitation during December–March tended to experience a decrease during April–July.

The period August–November shows a clear upward trend in precipitation (Fig. 2.21). Precipitation during the last 30 years has been about 20 percent greater than what occurred during the first half of the twentieth century. The greatest increase has occurred in the eastern and southern parts of Texas, the opposite of the spatial pattern of long-term trends in December–March.

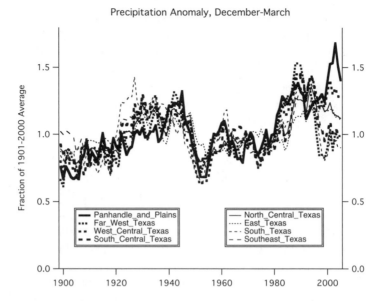

FIGURE 2.19. Nine-year running mean of average values of December–March USHCN Version 2 precipitation, as a proportion of the long-term mean, in each of eight Texas climate regions.

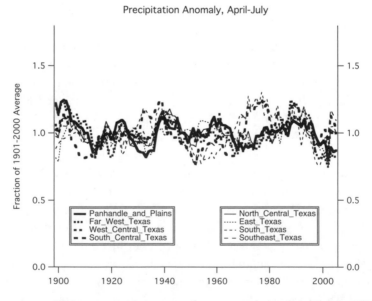

FIGURE 2.20. Nine-year running mean of average values of April–July USHCN Version 2 precipitation, as a proportion of the long-term mean, in each of eight Texas climate regions.

Overall, long-term precipitation trends range from increases of about 2 percent per century in the Panhandle and Plains and Far West Texas to about 20 percent more per century in South and Southeast Texas. A sense of the spatial pattern of long-term trends is given by data from individual USHCN stations in Texas and surrounding states (Fig. 2.22): the greatest increase in precipitation has occurred along a north-south corridor from South Texas to Oklahoma.

Future Precipitation Trends

The precipitation trends in Figure 2.22 are not consistent with expectations under global warming scenarios. The RCPM results for precipitation projections indicate a decline in precipitation toward the middle of the twenty-first century (Fig. 2.23). The median rate of decline in the models (about 10 percent per century) is smaller than the observed rate of increase over the past century. There is also considerable disagreement among models over whether precipitation will increase or decrease before the middle of the twenty-first century.

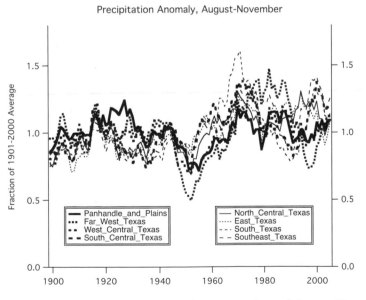

FIGURE 2.21. Nine-year running mean of average values of August–November USHCN Version 2 precipitation, as a proportion of the long-term mean, in each of eight Texas climate regions.

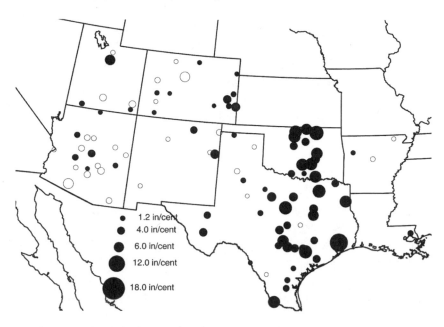

FIGURE 2.22. Precipitation trends at long-term USHCN Version 2 stations, in inches per century. Solid circles represent increases; open circles represent decreases.

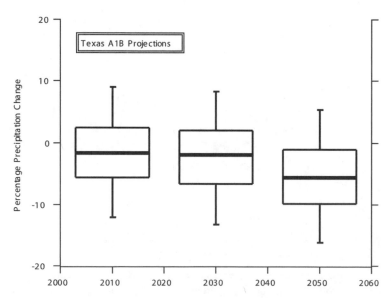

FIGURE 2.23. Statistical estimates of ranges of future accumulated precipitation in Texas relative to 1980–1999 baseline, according to Bayesian analysis of global climate models driven by the A1B forcing scenario. Each estimate is for a 20-year average. RCPM data and analyses are © 2006 University Corporation for Atmospheric Research (All Rights Reserved).

Although the climate models tend to agree on the overall global patterns of precipitation changes, they produce a wide range of precipitation patterns on the scale of Texas itself, such that there is no portion of the state that is more susceptible to declining precipitation in the model projections than any other.

Less research has been directed toward understanding discrepancies between precipitation trends and model projections than toward discrepancies between temperature trends and model projections. In part, this may be because model simulations driven by observed sea surface temperature variations have not yet been able to reproduce the observed substantial variations in precipitation patterns over the past century. Thus, a wide range of factors may have contributed to the observed precipitation increase.

An examination of these factors lends some indication of whether the increasing trend should be expected to continue. One possibility is that the increased precipitation is a direct response to global warming, albeit one that models have so far been unable to simulate. If so, precipitation would be likely to continue to increase. Another possibility is that the precipitation increase is a response to increased pollution from particulates and aerosols, either directly through changes in the precipitation formation process or indirectly through changes in weather patterns. If so, the trend should reverse itself as particulate pollution decreases. Another possibility is that the increased precipitation is a response to changes in land cover, such as irrigation, surface water storage, and urbanization. An irrigation mechanism seems unlikely, because the precipitation trend is absent precisely during the main agricultural growing season when irrigation is most common. Enhancement of precipitation near Houston, Texas, has been attributed to Houston urbanization (Burian and Shepherd 2005). A final possibility is that, like surface temperatures, precipitation changes are largely attributable to patterns of sea surface temperature changes. If so, future trends depend on whether the sea surface temperature changes themselves are primarily due to natural variability (in which case the precipitation trend would reverse itself) or primarily due to anthropogenic effects (in which case the precipitation trend would continue).

Absent any solid knowledge of the cause of the precipitation trends in Texas over the past century, it is impossible to say with confidence what the trends should be over the next half-century. Unlike temperature, where the expected temperature effect is much larger than natural variability, the model consensus for precipitation projections has a change by mid-century that is smaller than past observed rates of change. The models suggest that the odds are tilted in favor of a precipitation decrease, but the models' ability to simulate precipitation changes accurately over a relatively small area such as Texas has not been demonstrated. Within the realm of possibility are even more rapid increases in precipitation or a catastrophic reversal of the precipitation trend. About the only thing that can be said with confidence is the following: it is not appropriate to assume that the precipitation climatology has been or will be steady over time.

COMBINED CHANGES IN TEMPERATURE AND PRECIPITATION

Much of the Texas natural environment is highly sensitive to soil moisture and water availability, and the Texas human environment (both urban and agricultural) is becoming increasingly sensitive to water supply and demand. These parameters all depend on the effect of simultaneous changes of temperature and precipitation.

The sharp seasonal contrasts in Texas extend to the relationship between temperature and precipitation. During the wintertime in most of Texas, increased precipitation is often accompanied by neutral or increased temperatures, all associated with increased transport of heat and moisture from the Gulf of Mexico. Typically, more rain and snow falls than can evaporate, so even if temperatures are unusually warm there is an ample supply of environmental moisture. This lends a relative stability to cool-season water supplies. West Texas is quite a bit drier than the rest of the state during the winter, and increased moisture is typically accompanied by cool weather.

In the summertime, cooler weather and increased rainfall go hand in hand. Summertime surface temperatures are strongly affected by variations in soil moisture and cloud cover. Dry summertime conditions are also typically hot summertime conditions, so a lack of water supply (rainfall) is typically aggravated by an increase in water demand (evaporation).

The projected increase in temperatures across Texas implies warmer conditions when it is sunny as well as warmer conditions when it is cloudy. Thus, for a given amount of precipitation, water demand and evaporation will increase and water supply (in the form of reservoir storage, stream flow, and aquifer recharge) will decrease. So, even though precipitation projections are uncertain, it is likely that environmental and human water systems will be under increased stress due to changes in both water supply and water demand. If precipitation actually increases, the gains in precipitation will be partially or wholly offset by increased evaporation and evapotranspiration. If precipitation decreases, the decline in precipitation will be compounded by the increase in evaporation to yield an even greater reduction in water supply and an even greater increase in water demand.

CHANGES IN SEVERE AND HIGH-IMPACT WEATHER

The challenge of predicting future changes in severe and high-impact weather is threefold. First, such weather events are by their nature rare, so it is difficult to tell from observations within a single geographical area whether a true long-term trend is present. Second, observations of such weather events are much less consistent over time than conventional climate parameters such as temperature and rainfall, making it difficult to distinguish changes caused by the atmosphere from changes caused by observing and reporting practices. Third, all such weather events occur on a much smaller scale than climate models can directly simulate, requiring indirect techniques to infer changes in severe and high-impact weather from the climate model output.

One example of such a weather phenomenon is the tornado. A tornado is much too small to be simulated by present-day short-range forecasting models, let alone by global climate models. The large-scale meteorological conditions conducive to tornado formation are understood only in general terms, with much sensitivity to low-level frontal structure and other such atmospheric features, so no studies have been able to infer tornado frequencies from global climate simulations. Finally, the tornado records themselves are spatially inhomogeneous and unreliable, particularly for the weak tornadoes that are most common (Brooks and Doswell 2001).

It is worth noting that strong tornadoes in the United States show no detectable long-term trends (Brooks and Doswell 2001), and neither do most other types of severe weather, with the exception of heavy rainfall events (Balling and Cerveny 2003). As water vapor content increases with global temperatures, the amount of precipitation produced per storm should increase as well, and such a change has been observed. Inferences of changes in the likelihood of other types of severe weather under global warming scenarios are just beginning; initial results suggest an increase (Trapp et al. 2007).

Hurricane counts in the Atlantic fluctuate considerably from year to year and decade to decade. The quality of the Atlantic hurricane database prior to the satellite and aircraft reconnaissance eras is under active scientific debate; whether the recent increase in Atlantic hurricane activity is attributable to global warming or temporary sea surface temperature variations is also a matter of debate. Since global climate models cannot simulate hurricanes directly, recent studies have used a variety of techniques to embed direct simulations of hurricanes within a climate model simulation. Results here too have been mixed, and the balance of evidence suggests a future with fewer but more intense hurricanes (Knutson et al. 2010).

Despite this uncertainty, one can confidently predict that the likelihood of a major hurricane striking the Texas coast will increase over the coming half-century. The basis for this forecast is the following: landfalling hurricanes in the United States show insignificant trends over the past century or more (Wang and Lee 2008), yet, for no obvious reason, Texas has had fewer than its share of major damaging hurricanes over the past half-century. According to Blake et al. (2007), of the 10 strongest hurricanes to strike Texas in the past 150 years, only one (Rita) occurred within the past 25 years, and even that storm made landfall in Louisiana. Of the 25 storms estimated to be capable of causing the most damage, based on present-day population and infrastructure, 4 have hit Texas, but none since 1961. The return period for a major hurricane to strike Houston is 21 years, yet no major hurricane has hit the city since 1941. Even Ike, which caused widespread damage in the Houston area in 2008, did not reach the major hurricane threshold for wind speed at landfall. So, although it is not quite correct to say that Texas is overdue for more major hurricanes, it is correct to say that the expected frequency of very damaging hurricanes during the next few decades is higher than the actual frequency over the past few decades.

The last form of high-impact weather to be considered in this chapter is drought. Drought is expected to increase in general worldwide, because of the increase in temperatures and the trend toward concentration of rainfall into events of shorter duration. In Texas, temperatures are likely to rise, and future precipitation trends are difficult to call, so it is likely that drought frequency and severity will increase in Texas. If temperatures rise and precipitation decreases, as projected by climate models, Texas would begin to see droughts in the middle of the twenty-first century that are as bad as or worse than those in the beginning or middle of the twentieth century.

CONCLUDING REMARKS

Climate projection is inherently more difficult for a small subarea of the globe than for the globe as a whole. This is particularly true for Texas, where the nature of the climate changes throughout the year. The wide range of weather conditions that have significant impacts on Texas further complicates matters.

Partially because of this complexity, climate variations over the past century in Texas do not correspond to changes expected from global warming, according to present-day climate models. Local temperature changes due to global warming are likely to soon become strong enough to overwhelm natural variability, leading to temperatures in the neighborhood of 4°F warmer than recent decades by the middle of this century. In the case of precipitation, observed variations over the past century are larger than most future climate projections of precipitation change by mid-century, and are also unexplained. Thus, it cannot be said whether future precipitation will be more or less than present-day precipitation in Texas.

REFERENCES

Balling, R. C. Jr., and R. S. Cerveny, 2003. Compilation and Discussion of Trends in Severe Storms in the United States: Popular Perception v. Climate Reality. *Natural Hazards* 29:103–112.

Blake, E. S., E. N. Rappaport, and C. W. Landsea, 2007. The Deadliest, costliest, and most intense United States tropical cyclones from 1851 to 2006 (and other frequently requested hurricane facts). NOAA Technical Memorandum NWS TPC-5.

Brooks, H., and C. A. Doswell III, 2001. Some Aspects of the International Climatology of Tornadoes by Damage Classification. *Atmospheric Research* 56:191–201.

Burian, S. J., and J. M. Shepherd, 2005. Effects of Urbanization on the Diurnal Rainfall Pattern in Houston. *Hydrological Processes: Special Issue on Rainfall and Hydrological Processes* 19:1089–1103.

Easterling, D. R., T. R. Karl, E. H. Mason, P. Y. Hughes, and D. P. Bowman, 1996. U.S. Historical Climatology Network Monthly Temperature and Precipitation Data. ORNL/CDIAC-87, NDP-019/R3. Carbon Dioxide Information Analysis Center, Oak Ridge National Laboratory, U.S. Department of Energy, Oak Ridge, Tennessee.

Knutson, T. R., T. L. Delworth, K. W. Dixon, I. M. Held, J. Lu, V. Ramaswamy, M. D. Schwarzkopf, G. Stenchikov, and R. J. Stouffer, 2006. Assessment of Twentieth-century Regional Surface Temperature Trends Using the GFDL CM2 Coupled Models. *Journal of Climate* 19:1624–1651.

Knutson, T. R., J. L. McBride, J. Chan, K. Emanuel, G. Holland, C. Landsea, I. Held, J. P. Kossin, A. K. Srivastava, and M. Sugi, 2010. Tropical Cyclones and Climate Change. *Nature Geoscience.* DOI: 10.1038/ngeo779.

National Center for Atmospheric Research, 2008. RCPM: Regional Climate-Change Projections from Multi-Model Ensembles. http://rcpm.ucar.edu/, accessed May 21, 2008.

National Climatic Data Center, 2008. The USHCN Version 2 Serial Monthly Dataset. http://www.ncdc.noaa.gov/oa/climate/research/ushcn/, accessed May 20, 2008.

National Severe Storms Laboratory, 2003. Severe Thunderstorm Climatology. http://www.nssl.noaa.gov/hazard/, accessed May 19, 2008.

Nielsen-Gammon, J. W., F. Zhang, A. M. Odins, and B. Myoung, 2005. Extreme Rainfall in Texas: Patterns and Predictability. *Physical Geography* 26:340–364.

Pielke, R. Sr., J. Nielsen-Gammon, C. Davey, J. Angel, O. Bliss, N. Doesken, M. Cai, S. Fall, D. Niyogi, K. Gallo, R. Hale, K. G. Hubbard, X. Lin, H. Li, and S. Raman,

2007. Documentation of Uncertainties and Bias Associated with Surface Temperature Measurement Sites for Climate Change Assessment. *Bulletin of the American Meteorological Society* 88:913–928. DOI: 10.1175/BAMS-88-6-913.

Robinson, W. A., R. Reudy, and J. E. Hansen, 2002. General Circulation Model Simulations of Recent Cooling in the East-central United States. *Journal of Geophysical Research, Atmosphere* 107:D24, 4748. DOI: 10.1029/2001JD001577.

Tebaldi, C., R. L. Smith, D. Nychka, and L. O. Mearns, 2005. Quantifying Uncertainty in Projections of Regional Climate Change: A Bayesian Approach to the Analysis of Multimodel Ensembles. *Journal of Climate* 18:1524–1540.

Trapp, R. J., N. S. Diffenbaugh, H. E. Brooks, M. E. Baldwin, E. D. Robinson, and J. S. Pal, 2007. Changes in Severe Thunderstorm Environment Frequency during the 21st Century Caused by Anthropogenically Enhanced Global Radiative Forcing. *Proceedings of the National Academy of Sciences* 104:19719-19723. DOI: 10.1073/pnas.0705494194.

Wang, C., and S.-K. Lee, 2008. Global Warming and United States Landfalling Hurricanes. *Geophysical Research Letters* 35:L02708. DOI: 10.1029/2007GL032396.

Zhang, X. B., F. W. Zwiers, G. C. Heegerl, F. H. Lambert, N. P. Gillett, S. Solomon, P. A. Stott, and T. Nozawa, 2007. Detection of Human Influence on Twentieth-century Precipitation Trends. *Nature* 448:461–464. DOI: 10.1038/nature06025.

CHAPTER 3

Water Resources and Water Supply
George H. Ward

Of all the elements of the Texas economy, society, and environment considered in this book, water is most closely coupled with climate. It is also the quintessential limiting factor for human development of the state. Simply put, "the dominant feature of Texas is water, or rather, its scarcity" (Fehrenbach 1983). The present chapter focuses on availability of fresh water, that is, water with sufficiently low dissolved solids that it can be used for human and animal consumption, for the various agricultural and industrial enterprises, and for the wide suite of biological processes that require water of this quality. Specifically, this chapter addresses the question of whether there is a realistic potential for global warming, and associated climatological changes, to impact the availability of water in the state, and it seeks an answer by a rudimentary accounting of the sources and uses of water in the state and their responses to meteorological variables.

TEXAS HYDROLOGY

The ultimate source of fresh water in the state is precipitation, almost entirely rainfall. To understand the challenge of water management in the state and to anticipate the effect that a greenhouse-warmed climate may have, it is necessary to trace the disposition of water after its delivery as rainfall onto the landscape of the state, as illustrated in Figure 3.1. Rainfall impingent on the surface immediately begins infiltrating into the soil. If the rainfall rate exceeds the infiltration rate, the excess ponds on the surface. Once this ponding suffices to establish continuity across the surface, the excess water can flow downslope. This downslope flow, called runoff, becomes organized into the network of drainageways incised into the landscape surface that collect and convey water (streams and rivers). Over a longer time period, some of the water that infiltrates the near-surface layer of the soil is evaporated, some is taken up by plant roots and ultimately transpired back to the atmosphere, some moves laterally and emerges down-gradient at the surface (as seeps and interflow), and some percolates downward into aquifers, deep water-bearing formations.

There are two components of this hydrological network in which water volume is sufficiently and reliably concentrated to serve as managed water

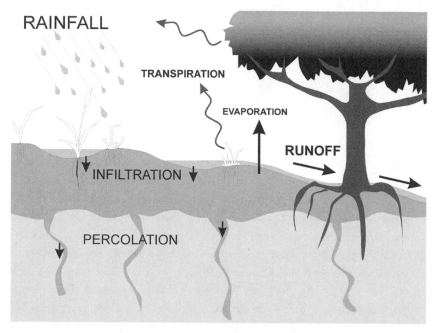

FIGURE 3.1. Water transfers in the landscape, above and below ground, including runoff to surface water and percolation to groundwater.

supply; the first is the river systems of the state and the second is the aquifers, referred to respectively as surface water and groundwater. Surface water encompasses all of the watercourses on the surface of the land in which water flows or is accumulated, including rivers and their tributaries, lakes, and reservoirs. A river basin is a river's watershed (technically, the watershed at the river mouth), the geographical region that supplies runoff to the stream channels of the river system. (The principal rivers of Texas and their associated basins are shown in Figure 3.2.) Groundwater is the water contained in permeable rock formations, of which 9 major and 20 minor aquifers provide important water-supply sources in Texas (Fig. 3.3). The influx of water to an aquifer, percolating down from the surface, is referred to as recharge. The extent to which an aquifer is recharged is strongly dependent on the overlying strata, the nature of the land surface and soils, the hydroclimatology of the recharge area, and the geological properties of the formation itself.

In Texas, runoff is usually produced during and immediately after thunderstorm events, which are characteristically brief and intense but infrequent. Rainfall, and therefore runoff, exhibit long-term patterns in time and space that emerge when data are appropriately averaged. The frequency and

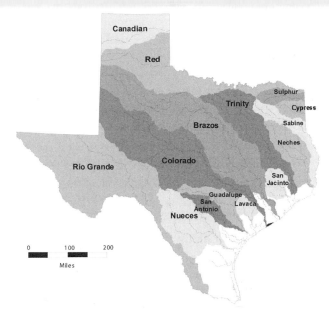

FIGURE 3.2. Principal rivers of Texas and their basins

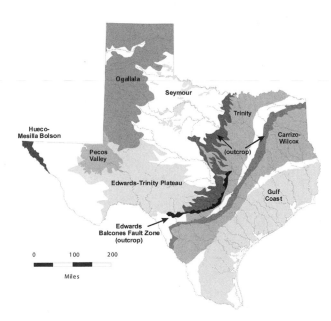

FIGURE 3.3. Principal aquifers of Texas, with rivers superposed. An outcrop is that part of a water-bearing rock layer that appears at the land surface.

intensity of storm events have a definite seasonality, with maxima in spring and fall in most areas of the state. There is also a pronounced variation in annual rainfall across the state. As described in Chapter 2, annual rainfall declines precipitously from east to west across the state, by a factor of 6 to 7 (depending on the time period of averaging). The proportion of rainfall that appears as runoff (the runoff-to-rainfall ratio) similarly declines by more than an order of magnitude (Fig. 3.4). The products of these, runoff and stream flow, decline by nearly two orders of magnitude from east to west. The contrast between the geographical distribution of surface water in the state and the geographical distribution of the demand for water is a defining feature of the challenge of water management in Texas.

The convective nature of precipitation in Texas has profound conse-quences for water availability and water use. Texas does not accumulate a

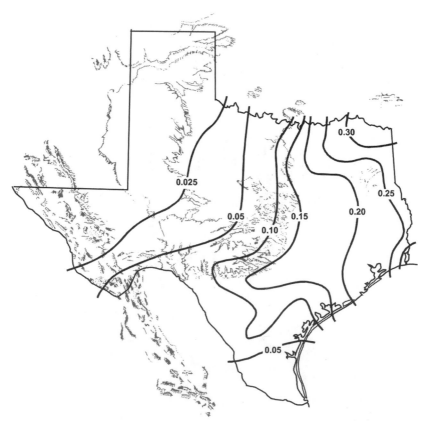

FIGURE 3.4. Contours of the ratio of runoff to rainfall across Texas. The northeast corner of the state has the highest proportion of runoff (0.30), about 10 times higher than the proportion in the Panhandle and Far West Texas.

winter mass of snow that slowly melts in the spring, feeding lakes and rivers, nor is Texas subjected to stationary stratiform weather systems that sustain moderate-intensity but prolonged rainfall over large regions. Rather, Texas' precipitation is derived from deep-convecting storm systems and individual thunderstorm cells. Precipitation, consequently, is locally intense but short-lived; rainfall is intense when it occurs, but of brief duration. As this precipitation falls on the surface and is organized into stream and river channels, the corresponding river flows are *flashy*, with pronounced peaks in flow and rapid rise and recession. Flow in a Texas stream may be characterized as a small base flow, on which are superposed these occasional storm hydrographs, with seasonal variations in frequency and intensity, often separated by long periods of low flow.

This variability means that river flows are not dependable as a water source, which poses a major problem in using surface water for water supply. Most of the time in most of the state, there will be too little river flow to meet the needs of water supply. This problem is dealt with through the construction of dams, which impound reservoirs that capture some of the higher flows of the rivers and hold this water for use during dry periods. The reservoir is the cornerstone of surface-water management in Texas. Nowhere in the state is there any significant use of uncontrolled, run-of-the-river flow as water supply. According to the Texas Water Development Board (TWDB), there are presently 195 major reservoirs in Texas (a capacity exceeding 5,000 acre-feet is considered major). They have a combined storage capacity of 38.7 million acre-feet (maf) allocated for conservation (i.e., water supply; many have additional capacity allocated for flood control, but that is not used for conservation storage and therefore is not considered here). In addition, there are some 1,500 flood and erosion control reservoirs (see, for example, the small subwatershed of the Bosque in Figure 3.5) and perhaps 300,000 smaller reservoirs, private lakes, farm ponds, and stock tanks. Clearly, the surface of Texas is extensively plumbed. (See Ward 2004 for information about the history of water-resource development in the state, as well as more discussion of state hydroclimatology.)

WATER BUDGETING

A valuable technique for analyzing the availability of water is a water budget. As the name implies, this is an accounting of water transfers, analogous in many respects to financial transactions, in which the sources and disposition of money are identified. A water budget similarly analyzes the different compartments or accounts of water and the transfers among them. A water budget is carried out for some well-defined region in space

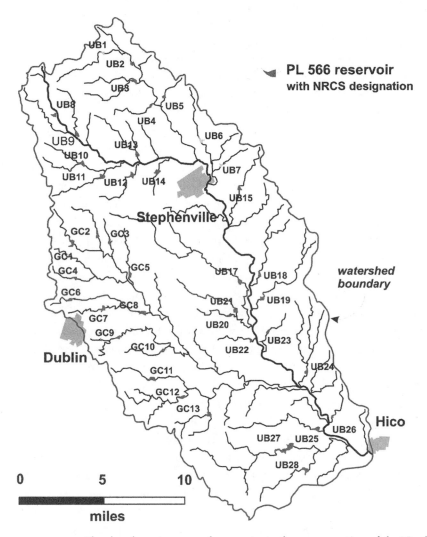

FIGURE 3.5. Flood and erosion control reservoirs in the upper portion of the North Bosque River watershed, above Hico. These are designated by the Natural Resources Conservation Service (formerly Soil Conservation Service) under Public Law 566 and known as PL566 reservoirs.

and integrated over some definite period of time. Clearly, water budgets can be performed for smaller and smaller units (e.g., a river basin, a county, a city, or even an agricultural field or house and lot), depending on the purposes of the analysis. In the present context, we will present water budgets closed over the entire state or in large subregions of the state. We will also average the water budget over periods of many years. For the hydroclimatological components, this period will typically be 30 years, the conventional averaging period of a climatological norm.

Unlike a financial budget, there is little documentation for transfers of water, nor are there records of the balances in the different water accounts. These must be discovered by direct measurement, by inference from meteorological or hydrological principles, or by posing and testing alternative assumptions. Closing a water budget can become a challenging scientific endeavor, and the results are frequently unexpected.

An approximate water budget for Texas (Fig. 3.6) would begin by formalizing the transfers of water illustrated in Figure 3.1. Although the results of this water budget are presented for the entire state, they were not computed this way. Instead, the state was subdivided into four large regions (Fig. 3.7), representing broadly distinct hydroclimatologies, and a water budget analogous to the one in Figure 3.6 was developed for each region. These were then combined to determine the statewide budget. Though each of these regions is far from homogeneous in climate, vegetation, and hydrology, as a group they generally characterize the range in hydroclimatology of the state.

The details of the water-budget methodology follow those presented in Ward (1993), with the following exceptions: (1) more recent data were employed, as described below; (2) runoff was directly computed from rainfall using the regression relations of the U.S. Geological Survey (USGS; Lanning-Rush 2000); (3) data on electrical generation and associated water uses were taken from King et al. (2008), rather than the Public Utilities Commission; (4) the budget includes computation of runoff from adjacent states (using the nearest regional USGS relation); (5) inflow from Mexico into the Rio Grande, which was not considered in the Ward (1993) budget, is a specified input variable; (6) recharge is based on TWDB data (Muller and Price 1979), statistically related to rainfall (rather than to runoff); (7) rather than incorporate measured stream flow into the elements of the water budget, as done in Ward (1993), the present approach directly computed all elements of the budget (other than water uses and the inflow from Mexico, which are inputs) and validates the predicted stream flow against USGS measurements. As in Ward (1993), all of the reservoirs within each region are lumped as a single element at the site of the lowermost dam in the basin. Thus, their combined effect is modeled as a reservoir with surface area and conservation storage equal to the totals for the region.

Meteorological data for this water budget were taken from the National Climatic Data Center 1971–2000 climatological norms for divisions of Texas, New Mexico, Oklahoma, Arkansas, and Louisiana. Lake evaporation

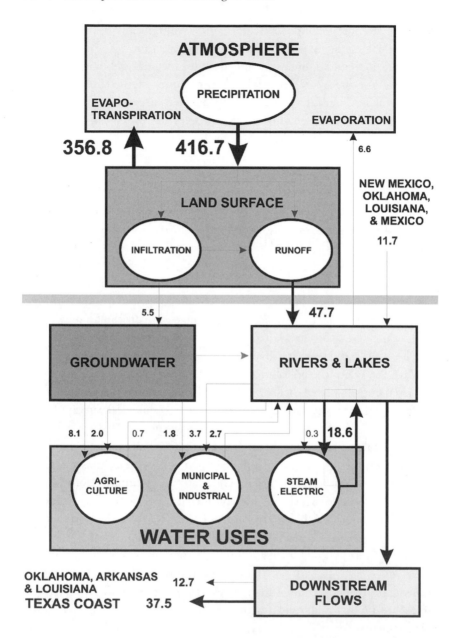

FIGURE 3.6. Water budget for Texas, under Present Normal conditions. The transfer of water from one compartment to another is indicated by an arrow, and the magnitude of that transfer is shown by the adjacent number, in millions of acre-feet per year. The units are less important for this discussion than the relative size of the transfers. To emphasize this, the larger transfers are shown by larger and bolder arrows and numerals. Based on 1971–2000 average meteorology and hydrology, and 2000 data on water uses.

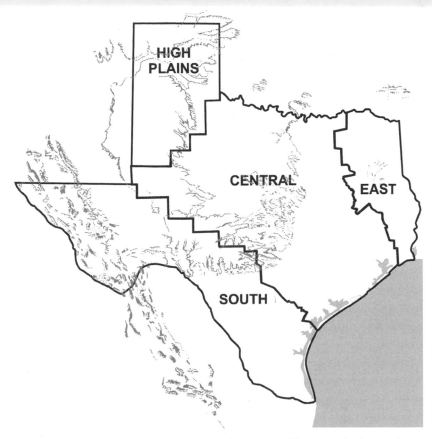

FIGURE 3.7. Hydroclimatological regions of Texas used for water-budget analyses. The names of the regions apply only in the present context and only poorly approximate the usual geographic terminologies.

data compiled monthly and aggregated by the TWDB were averaged over the same period. Streamflow data of the U.S. Geological Survey averaged over the 1971–2000 period were used to validate the computed runoff for each of the regions. Unfortunately, we do not have a 30-year record of water use to average in the same way, so instead, water-use data for the year 2000 from the TWDB were used. Reservoir status as of the year 2000 was used to characterize area and volume of the conservation and power reservoirs of the state, based on data compilations of the TWDB. The conditions depicted in Figure 3.6, which we refer to as Present Normal for simplicity, are therefore a sort of mongrel: more-or-less current water uses imposed on a 30-year average hydroclimatology.

WATER USE

In discussing human water use, one must distinguish between withdrawal and consumption. *Withdrawal* means the removal of water from a surface water or groundwater resource for some purpose (the term *diversion* is synonymous). Once water is withdrawn and used, it may be returned to the water resource as a *return flow*. For example, municipal water use entails a withdrawal for water supply and a return flow of treated wastewater. Of course, the quality of the two may be very different, and contamination can delimit further use of the water. This discussion views water supply in Texas from a broad perspective with the focus on water quantity; it therefore disregards water quality issues and considers a return flow to be a credit to the available water resource. (This will be qualified further in the discussion of specific uses.) *Consumption* means that the water is permanently removed from the water resource. The difference between withdrawal and return flow for a specific use is the water consumption of that use. Water that is consumed generally is eventually emitted back to the atmosphere in some form and may find its way back into the water supply through precipitation, but the accounting of a water budget disregards such consumed water for the simple reason that it is no longer available for water supply.

Many human activities require the use of water in large volumes. For the present purposes, these are treated in the three broad categories of municipal and industrial uses, agriculture, and electric power generation. In the first category, municipal and industrial uses are combined because the two are generally associated with large population centers in Texas. It includes personal water consumption and domestic use, as well as water used as part of an industrial process. The principal return flow is treated wastewater. Advancements in wastewater treatment technology have resulted in significant improvements in the quality of such return flows.

The second broad category is agriculture, including crop and livestock production. The dominant water use in this category on a statewide basis is irrigation, which involves the diversion of large volumes of water for placement on soil. Some of this returns to the surface water system through runoff, but much of it ends up in the atmosphere, evaporating directly from the soil or being transpired as a byproduct of plant metabolism. (Some of Texas' most insidious water quality problems are associated with agricultural return flows, which contain elevated nutrient and pesticide levels.)

The third category of water use is electric power generation, or, more specifically, steam-electric generation. The dominant form of power generation in Texas, steam-electric generation exploits the thermodynamics of evaporation and condensation of pure water to create high-pressure steam that spins turbines. In the process, this steam is recondensed by the use of large volumes of cooling water drawn from the surface water supply. Cooling water returned to the surface resource is altered only by its increase in temperature. This excess heat is quickly dissipated to the atmosphere, and the water can be reused for cooling. In fact, many of the steam-electric plants in Texas are situated on a dedicated reservoir, continuously reusing the water for cooling.

There are, of course, other uses of water, such as mining and recreation, but the three categories listed here dominate water use in Texas. These categories refer strictly to water used by people and their enterprises. The maintenance of natural ecosystems, notably the streams and rivers and lakes in the interior of the state and the estuaries along the Texas coast, requires a dependable flow of water. Maintaining the integrity of these ecosystems is also important for the economy of the state. That is why the water budget includes a category for downstream flows, which ultimately leave the state, either as a transfer into neighboring states or as a flow into the Gulf of Mexico.

An inspection of the water budget of Figure 3.6, crude as it is, reveals several perhaps surprising facts about water use in Texas. First, of the total runoff flowing into the reservoirs, less than 10 percent is removed for all human uses (a net 2.6 maf per year withdrawn, from 47.7 maf per year total runoff). This is mainly because the reservoirs are able to capture only a fraction of the higher flows of the rivers; the rest flows through (or over) the dams and, ultimately, to the Gulf of Mexico. Next, municipal (and industrial) surface withdrawals are offset by a return flow of 70 percent of the volume withdrawn. Such a high proportion is due to the additional return flows from groundwater withdrawal. Also note the enormous withdrawal required for steam-electric generation (nearly 19 maf per year). This is, in fact, the largest single diverter of water in the state, but the water is almost entirely returned to the surface water resource (depicted in Figure 3.6 by the closed loop between "Rivers and Lakes" and "Steam Electric"). Thus, although the actual consumption of water for power generation is minuscule in comparison with total consumption by all uses, electric power requires the availability of a large volume of throughflow. The next largest category of withdrawal is for agricultural purposes. Agriculture is the dominant consumer of water in the state, with over three times the consumption of all other uses combined.

As part of the planning process, the Texas Water Development Board compiles projections of future water use for planning over several decades, based on population projections and scenarios of industry and agriculture development (TWDB 2007). This chapter employs the board's 2050 projected water uses, to correspond to the 50-year changed climate scenario (relative to a year 2000 baseline).

REGIONAL WATER BUDGETS, PRESENT AND PROJECTED NORMAL CONDITIONS

Over the 30-year averaging period of Figure 3.6, rainfall supplies roughly 417 maf of water to Texas per year. Less than 1 percent of this falls directly on lakes and rivers. The remainder of the influx to the lakes and rivers of the state must traverse the land surface. In the present context, the land surface is considered a processor of rainfall. Most of the precipitation falling on the land surface—nearly 90 percent statewide (Fig. 3.4)—does not make it to a surface water body but is intercepted by a complex of processes, the most important of which is evapotranspiration, the direct evaporation of water back into the atmosphere or uptake by plants and subsequent transpiration (Fig. 3.1). The efficiency of the runoff process is very low, therefore, and is a strong function of terrain, soils and near-surface geology, and vegetation. Antecedent conditions are particularly important; the runoff produced from a rainstorm is greatly reduced if the surface is desiccated, a characteristic of arid regions and of drought conditions when there are long intervals separating storm events.

Two of the most significant inferences to be drawn from this statewide water budget merit special comment. The first is that Texas, from a water supply viewpoint, is primarily a groundwater state. In terms of net withdrawals (i.e., with credit for agricultural, municipal, and industrial return flows), the groundwater withdrawal is nearly four times that from surface supplies. The second is that runoff is more than ample to meet the surface water requirements of the state, being well over three times the present total water demand. Both inferences are misleading. The problem is the geographical variation in the elements of the water budget. This is demonstrated by examining the separate water budgets for each of the four hydroclimatological regions in Figure 3.7, detailing those components driven by hydroclimatology, in this case the 1971–2000 means, but unaffected by human activities (Table 3.1), and the various water uses (from the TWDB 2000 data compilation), with the resulting downstream flows (Table 3.2). The flow to the Texas coast should be especially noted. A water budget analogous to Figure 3.6 could be diagrammed for each of these hydroclimatological regions using these data.

The High Plains region accounts for 70 percent of the groundwater withdrawal of the entire state, predominantly for agriculture. Since the region

TABLE 3.1. Regional hydroclimatology and water budget inflows, 1971–2000

	HYPERCLIMATOLOGICAL REGION[a] (MEAN TEMPERATURE, °F)				
	HIGH PLAINS	CENTRAL	EAST	SOUTH	STATE
Water Component, in millions of acre-feet per year	(59.0)	(65.4)	(65.4)	(67.1)	(64.9)
Precipitation	44.8	211.7	69.5	90.6	416.7
Evapotranspiration	43.7	182.9	43.8	86.3	356.8
Runoff	0.4	22.9	22.3	2.2	47.7
Recharge	0.5	2.4	1.2	1.3	5.5
Lake evaporation	0.1	3.5	2.2	0.8	6.6
Inflow from upstream					
New Mexico	0.1			0.4	0.5
Oklahoma		5.0	0.3		5.3
Louisiana			3.8		3.8
Mexico				2.2	2.2

[a] As defined in Figure 3.7

TABLE 3.2. Texas 2000 water uses by region and water-budget flows

	HYPERCLIMATOLOGICAL REGION[a]				
WATER SOURCE AND USE	HIGH PLAINS	CENTRAL	EAST	SOUTH	STATE
Surface water use					
municipal-industrial	0.09	2.54	0.65	0.39	3.7
agriculture	0.05	0.82	0.12	1.01	2.0
electric	0.02	0.24	0.06	0.00	0.3
Groundwater use					
municipal-industrial	0.13	1.27	0.21	0.21	1.8
agriculture	6.52	0.78	0.02	0.76	8.1
electric	0.02	0.09	0.01	0.02	0.1
Return flows					
municipal-industrial	0.08	1.99	0.48	0.21	2.7
agriculture	0.00	0.61	0.05	0.04	0.7
Steam-electric circulation	0.0	12.3	6.3	0.0	18.6
Downstream flow to					
Oklahoma	0.2				0.2
Arkansas			10.7		10.7
Louisiana			1.8		1.8
Texas coast		17.0	17.8	2.7	37.5

[a] As defined in Figure 3.7. Based on 1971–2000 regional hydroclimatology (Table 3.1), millions of acre-feet per year

has only 4 percent of the state's population, the other uses are minor at best. This enormous volume of groundwater is withdrawn from a single remarkable resource, the Ogallala formation. Its recharge is only about 5 percent of the withdrawal. Thus, from a water-supply viewpoint, the Ogallala is a finite resource that can only decline monotonically until depleted. There are also reservoirs in the High Plains, but runoff is quite low as a consequence of the arid climate. The reservoirs principally supply some of the municipal demands of the region. Clearly, the available runoff, even if totally captured (probably a physical impossibility), cannot begin to replace the Ogallala. The High Plains region is nearly isolated from the rest of the state's water budget, both as a source and a sink of water, but its inclusion in the state averages in Figure 3.6 greatly distorts the interpretation of that water budget. If the High Plains groundwater withdrawal is deleted from the budget, Texas appears to be about equally dependent on surface water and groundwater.

At the other extreme, the East Texas region (also sparsely populated, with about 8 percent of the state's population) has little groundwater usage but receives about half of the runoff in the state. Indeed, considering its area, this region has the greatest runoff rate in the state, by about a factor of 4, and is, therefore, the most water rich. Some of the water captured in reservoirs, and most of the electric power, is exported to the more populous regions of the state. The central concern, amounting to a fixation, of water-supply engineering in Texas for at least half a century has been devising a means of transporting water from this water-rich region to areas where demand exceeds supply (Ward 2004).

The Central Texas region contains the principal population centers of the state and most of its industrial capacity. The total volume of runoff in this region equals that of East Texas, but the runoff per unit area is considerably less. The groundwater withdrawal for agricultural and municipal use is dominated by one formation, the Edwards Aquifer, in south-central Texas. (The Edwards is the sole municipal water supply for the city of San Antonio, a fact that has erupted as a political issue in recent years because of competing demands for this water, both for agricultural irrigation farther west and for maintenance of spring and river flow.) Much like the Ogallala, the Edwards distorts the statewide picture of water supply. If it also is deleted from the water budget, Texas appears to be predominantly a surface-water state.

Finally, the South Texas region, which includes the vast semitropical areas of the Nueces and Rio Grande basins, is the most arid region of the state, with substantial areas of desert in the west. Groundwater usage for agriculture is dominated by withdrawals in the Winter Garden. Agricultural irrigation in the Rio Grande Valley is the dominant use of surface water, supplied mainly from the Rio Grande international reservoir system of Falcón and Amistad. The most significant feature of this region's water budget is the high ratio of reservoir supply to runoff. In other regions, this ratio is at least an order of magnitude lower (e.g., somewhat more than 10 percent in the Central Texas region). In the South Texas region, net surface water withdrawal is over 50 percent of runoff. Unlike the other regional water budgets, this one cannot be closed reliably for influxes from out of state. Most of the watershed is located in Mexico, and the majority of the flow of the Rio Grande is from the Rio Conchos. For the scenarios evaluated here, there are two candidates to represent the Mexican inflow: the average measured inflow based on gauged data of the International Boundary and Water Commission, which was 2.2 maf per year over the 1971–2000 period, and the minimum delivery mandated by the Treaty of 1944 at 0.35 maf per year. The former is used in Table 3.1.

According to the TWDB (2007), the Texas population in 2050 is projected to be nearly double the 2000 population (a factor of 1.97 to be precise), and the associated ratios of projected 2050 water uses relative to 2000 are municipal and industrial, 1.87 from surface water, 1.66 from groundwater; agriculture, 1.37 surface, 0.81 ground; and steam electric, 2.69 surface, 3.18 ground. These are substantial increases (except for the large decline in withdrawals for agriculture from the Ogallala). Regional data and the projected downstream flows from the water budget model are given in Table 3.3. Since this water budget is for average 1971–2000 conditions, the data of Table 3.1 apply to this scenario as well.

TABLE 3.3. Texas 2050 projected water uses by hydroclimatological region and resulting water budget flows

WATER SOURCE AND USE[a]	HYDROCLIMATOLOGICAL REGION[b]				
	HIGH PLAINS	CENTRAL	EAST	SOUTH	STATE
Surface water use					
municipal-industrial	0.11	4.90	1.06	0.81	6.9
agriculture	0.07	1.08	0.26	1.32	2.7
electric	0.03	0.65	0.18	0.03	0.9
Groundwater use					
municipal-industrial	0.16	2.25	0.28	0.33	3.0
agriculture	4.88	0.93	0.03	0.67	6.5
electric	0.04	0.25	0.11	0.05	0.5
Return flows					
municipal-industrial	0.09	3.73	0.75	0.39	5.0
agriculture	0.00	0.77	0.10	0.04	0.9
Steam-electric circulation	0.0	33.7	17.3	0.0	50.9
Downstream flow to					
Oklahoma	0.2				0.2
Arkansas			10.6		10.6
Louisiana			1.8		1.8
Texas coast		16.0	17.5	2.2	35.7

[a] From Texas Water Development Board (2007); based on 1971–2000 regional hydroclimatology (Table 3.1), millions of acre-feet per year

[b] As defined in Figure 3.7

DROUGHT

The focus of water planning and management in Texas is the drought. Water supply is predicated on what can be dependably available during the worst-case drought condition. The volume of water that can be removed from a water-supply source without failure during such drought conditions is its *firm yield*, and water in Texas has been historically allocated on the basis of firm yield. In present practice, the worst-case drought is defined as the most intense drought that has occurred during the period for which hydrometeorological data are available, known as the *drought of record*. From an engineering and water-management viewpoint, this has the considerable advantage of basing water management on real events—for which meteorological and hydrological measurements are available, and the effects of which have been actually demonstrated—rather than on a theoretical construct. Using a historical event for planning purposes is particularly advantageous from a policy standpoint. Water development is highly political, involving large public expenditures and the sacrifice of land for the creation of reservoirs and pipelines. It is far easier to argue such actions based on a real drought event rather than an abstract possibility.

The drought of record also presents an important disadvantage. By relying on it, we effectively assume that the most intense drought to have occurred in our short period of data collection (about 50 years for most rivers in Texas—at most, 100 years) is the very worst that nature can inflict on us. If our assumption is correct and we design our reservoirs and water supplies to accommodate that level of drought severity, then we will be all right. Given the variability of climate on time scales longer than our data record, however, this assumption is most likely untrustworthy.

Moreover, climate change may well increase the severity of extreme drought. For most of Texas, the drought of record was in the 1950s. During that decade, especially during the first six or seven years, rainfall events were sparse in time and limited in magnitude, leading to a cumulative surface-water shortage that was an economic disaster for the state (Ward 2004). The severity of a drought is measured by the combination of its intensity (the deficit below normal of rainfall) and its duration (the period over which the rainfall deficit is prolonged). The 1950s drought was especially severe because it was both intense and long, lasting nearly seven years.

Because of the central importance of the 1950s drought in Texas water planning, a separate water budget analysis was performed for this event. The hydrological and meteorological data compiled for 1950–1956, in Table 3.4, should be compared with the 1971–2000 normal conditions in Table

3.1. Most notable is the disproportionate 60 percent reduction in runoff, from 47.7 to 19.7 maf per year, compared with the more modest 25 percent reduction in rainfall. This is an example of how the landscape operates as an amplifier of changes in rainfall on the resulting runoff.

Although water-use data can be compiled for the 1950–1956 period, it is useful in the present context of estimating climate-change impacts to create a synthetic scenario in which current (2000) water uses are coupled with 1950s climatology. (Lake evaporation, for example, is much less for the 1950s because reservoir development was considerably less at the time. With 2000 reservoir development, this evaporation would increase from that shown in Table 3.1.) For this synthetic scenario, it is further assumed that the contribution of runoff from the Mexican portion of the Rio Grande watershed is limited to the minimum delivery of the Treaty of 1944.

POTENTIAL IMPACTS OF CLIMATE CHANGE

Hydrologists readily acknowledge that the ultimate controls on the surface-water budget are atmospheric. It is far easier, however, and in many respects more precise, to measure the manifestations of these controls in stream flow and to extend these results statistically than it is to measure and calculate the direct effect of atmospheric processes on hydrology. So long as climate is stable over the long term, this is certainly a valid and fruitful approach to hydrology. Indeed, the lower half of the water budget in Figure 3.6 depicts rather well the conventional hydrological model employed in engineering and water planning, as represented, for example, by the Water Availability Model of Texas (Wurbs et al. 2005). The information transfers from the

TABLE 3.4. Drought-of-record hydroclimatology and water budget inflows, 1950–1956

	HYDROCLIMATOLOGICAL REGION[a] (MEAN TEMPERATURE, °F)				
	HIGH PLAINS	CENTRAL	EAST	SOUTH	STATE
Water Component, in millions of acre-feet per year	(59.9)	(65.9)	(67.3)	(68.0)	(65.8)
Precipitation	30.6	170.1	52.4	64.0	317.1
Evapotranspiration	30.1	156.9	41.7	62.0	290.8
Runoff	0.1	9.5	9.5	0.6	19.7
Recharge	0.4	1.9	0.9	1.0	4.2
Lake evaporation	0.0	1.8	0.2	0.4	2.4
Inflow from upstream					
New Mexico	0.0			0.0	0.1
Oklahoma		2.3	0.2		2.4
Louisiana			2.6		2.6
Mexico				1.6	1.6

[a] As defined in Figure 3.7. Cf. Table 3.1

top half of the budget (the atmosphere and landscape components) into the lower half (surface-water and groundwater supply sources) are replaced by actual observations, suitably analyzed. Without the resource of long-term hydrological measurements, namely, lake evaporation rates, water-table elevations, and, most important, stream flows, accurate water planning would be impossible.

This method is, however, poorly equipped to address the impact of climate change, because this would require the explicit incorporation of dependencies on atmospheric conditions. In other words, the changed-climate analysis must restore the upper half of Figure 3.6 and specify, in addition, explicit dependencies on atmospheric variables. With the water budget constructed here, in which runoff is explicitly based on rainfall (Lanning-Rush 2000), a first approximation of potential climate effects can be explored. It assumes that climate effects in their entirety are driven by atmospheric temperature and precipitation, and it employs the same dependencies on these two variables used previously (Ward 1993), including the Dalton-law dependency of evaporation and steam-electric forced evaporation on air temperature, regional regressions of municipal and industrial water demand on air temperature, and electric power demand on air temperature. In addition, the increase in agricultural water demand with air temperature is modeled by the dependency used in Chapter 6, acknowledging that those data are limited to the Edwards Aquifer and specific crops, whereas the dependency is applied here statewide to the entirety of the agricultural demand.

Global warming is expected to increase the intensity of the global hydrologic cycle (e.g., Mitchell 1989; IPCC 2007). The impacts are quantified by application of global climate models and by reasoning from historical and prehistorical associations between climate indicators (summarized in Chapter 1). There is considerable uncertainty in any such quantitative climate forecast, especially as applied to the hydrological cycle, which is the end product of a chain of complex physical processes (and, therefore, in the global climate model, assumptions and approximations). This is exemplified in the range of predicted future temperature and precipitation changes generated by the 21 global climate models used for such simulation by the IPCC (2007).

Generally, an increase in temperature is indicated for the entire south-central United States, with reduced precipitation and drier soil conditions for the Texas area, but the range surrounding this prediction is considerable, especially for precipitation. In this chapter, the estimated 2050 climate for Texas is taken to be the predicted changes in air temperature and precipitation averaged over multiple global climate models, for scenario A1B obtained from IPCC (2007). (At the year 2050, the results for scenarios A1B and A2 are virtually indistinguishable.) The temperature change is about a +2°C (+3.6°F) increment, and the precipitation change is a 5 percent decrease (which, coincidentally—or perhaps not—are exactly the same estimates from the 1990 IPCC assessment used in my 1993 water budget for Texas and in the first edition of this book).

Based on these scenarios, water budgets have been analyzed for Texas and the four hydroclimatological regions, with and without additional effects of greenhouse warming, for normal and drought conditions with 2000 and 2050-projected water uses (Tables 3.5–3.9). The results are presented as a ratio (in percent) relative to the 2000 normal values. Results are included for the 50-year climate projection in concert with 2000 water uses, to facilitate separating the effect of future increased water use, driven by growth, from the effects of climate change. Although the postulated alterations in temperature and precipitation for the climate-change scenario are modest, their effect on the state's water resources is dramatic: a reduction of 17 percent in runoff and 26 percent in flows to the coast under normal conditions at year 2050 demands. Under drought conditions, the 2000 runoff and flows to the coast are reduced to 41 percent and 32 percent of normal, respectively (reductions of 59 and 68 percent from normal). The greenhouse-warmed 2050 conditions produce additional reductions of 15 percent in runoff (35 percent of normal) and 10 percent in flows to the coast (27 percent of normal).

That is only part of the impact. These water budgets do not indicate whether the water-use demands are in fact met. If reservoir volume is never driven to zero, then there is a surfeit of water above the total demand, and the water-use demands are provided. A negative surfeit indicates a deficit, or shortfall (the extent to which the surface water supply fails to meet the water-use demands). Under both normal and greenhouse-warmed conditions for both 2000 and 2050 levels of water use, there is not a significant shortfall in the state (Table 3.10). Under drought conditions, however, the situation is worse. The 2000 water uses will result in a substantial shortfall in the South region, primarily associated with the overallocation of the Lower Rio Grande, a statement that accords well with the experience of the recent drought in the Rio Grande Valley. Under 2050 water uses, shortfalls occur in each region except the East, and these are exacerbated under the greenhouse-warmed condition.

TABLE 3.5. Statewide water budget components for various scenarios

WATER COMPONENT[a]	SCENARIO			
	NORMAL CLIMATE	GREENHOUSE-WARMED	DROUGHT NORMAL	GREENHOUSE DROUGHT
Precipitation	100	95	76	72
Evapotranspiration	100	96	80	76
Runoff	100	83	41	35
Recharge	100	95	76	72
Lake evaporation	100	119	108	129

HUMAN WATER USE	WATER-USE SCENARIO YEAR						
	2000	2050	2000	2050	2000	2050	2050
Surface water							
municipal-industrial	100	187	106	199	101	190	201
agriculture	100	137	134	183	110	151	197
electric	100	269	131	353	107	288	375
Groundwater							
municipal-industrial	100	166	106	177	101	168	179
agriculture	100	81	134	108	110	89	116
electric	100	318	130	417	107	341	446
Return flows							
municipal-industrial	100	181	106	192	101	183	194
agriculture	100	131	134	175	110	144	188
Steam-electric circulation	100	275	104	286	101	278	289
Downstream flow to							
Other states	100	99	83	82	41	40	33
Texas coast	100	95	76	70	32	30	27

[a] Measured as a fraction (percent) of Present Normal conditions (Tables 3.1 and 3.2)

These water budgets assume the same levels of reservoir development in 2050 and 2000. This assumption and the indicated water-use shortfalls (Table 3.10) explain the paradoxical result in which the greenhouse scenario has more impact to flows to the coast under normal conditions (26 percent) than under drought conditions (10 percent). The assumption is unrealistic, because additional surface supplies will be mandated by the doubled population. While future reservoir development cannot be projected with any confidence (it will most likely entail multiple reservoir sites with extensive interbasin transfers), we can impose on the water budget the least-development assumption, in which new supplies are created to meet only and exactly the water-use shortfalls of Table 3.10 (meaning that the High Plains shortfalls are met in the East region). Under this scenario, the resulting flows to the Texas coast are greatly reduced (Table 3.11). At 2050 projected water uses under drought conditions, the effect of greenhouse climate change is to reduce flows to the coast by an additional 42 percent statewide, to a level 15 percent of the 2000 normal.

TABLE 3.6. High Plains region water budget components for various scenarios

WATER COMPONENT[a]	NORMAL CLIMATE	GREENHOUSE- WARMED	DROUGHT NORMAL	GREENHOUSE DROUGHT
	SCENARIO			
Precipitation	100	95	68	65
Evapotranspiration	100	95	69	65
Runoff	100	84	29	24
Recharge	100	95	68	65
Lake evaporation	100	121	106	128

HUMAN WATER USE	2000	2050	2000	2050	2000	2050	2050
	WATER-USE SCENARIO YEAR						
Surface water							
municipal-industrial	100	122	107	131	102	125	133
agriculture	100	136	134	181	110	149	195
electric	100	135	128	173	107	145	183
Groundwater							
municipal-industrial	100	126	107	135	102	128	137
agriculture	100	75	134	100	110	82	108
electric	100	206	128	263	107	220	278
Return flows							
municipal-industrial	100	125	107	133	102	127	136
Downstream flow to							
Other states	100	103	111	90	40	43	40

[a] Measured as a fraction (percent) of Present Normal conditions

DISCUSSION AND CONCLUSIONS

The water budgets presented here are rudimentary and approximate. They are proffered as a means of addressing the question stated at the outset, whether the Texas water supply is potentially vulnerable to climate changes on the order of those projected for a greenhouse-warmed scenario. The answer is clearly affirmative. Taking flows to the coast as a measure of river-basin impact, the net effect statewide of the assumed greenhouse climate change, a 3.6°F increase in air temperature and a 5 percent decrease in precipitation, is to reduce these flows by about 25 percent under normal conditions and by about 45 percent under drought conditions, relative to the already reduced flows under 2050-projected water-use demands (Table 3.11). The 2050 projected flows to the coast are 70 percent of the 2000 normal values under normal conditions with the effect of a greenhouse climate imposed, and 15 percent of 2000 normal under drought conditions. In general, the effect of climate on water demands and watershed processing of rainfall is to amplify the changed-climate signal, because the causal connections are nonlinear and reinforcing.

TABLE 3.7. East region water budget components for various scenarios

	SCENARIO						
WATER COMPONENT[a]	NORMAL CLIMATE		GREENHOUSE-WARMED		DROUGHT NORMAL		GREENHOUSE DROUGHT
Precipitation	100		95		75		72
Evapotranspiration	100		96		91		87
Runoff	100		86		43		37
Recharge	100		95		75		72
Lake evaporation	100		120		100		120
	WATER-USE SCENARIO YEAR						
HUMAN WATER USE	2000	2050	2000	2050	2000	2050	2050
Surface water							
municipal-industrial	100	164	103	168	102	166	171
agriculture	100	230	134	307	110	253	330
electric	100	275	138	379	107	294	438
Groundwater							
municipal-industrial	100	134	103	138	102	136	140
agriculture	100	133	134	177	110	146	191
electric	100	815	138	1125	107	873	1299
Return flows							
municipal-industrial	100	157	103	161	102	159	163
agriculture	100	213	134	285	110	234	306
Steam-electric circulation	100	275	104	286	102	281	289
Downstream flow to							
Other states	100	100	86	86	68	68	62
Texas coast	100	98	82	80	42	40	30

[a] Measured as a fraction (percent) of Present Normal conditions

It is important to observe that nearly every assumption or approximation in this argument is imposed to minimize the estimated effect of climate change. The first such assumption is the estimated changed-climate condition itself, a uniform statewide increase in average temperature and decrease in average precipitation that is a long-term average (10–20 years centered on the 2050 forecast) of model results averaged over 21 models of varying skill in depicting meteorology in the Texas area. Most of the model simulations indicate greater temperature rises and less rainfall in the interior and western areas of the south-central United States, effects that become more exaggerated with distance south into Mexico (IPCC 2007). This would imply less runoff into those reservoirs in the upland reaches of Texas basins, and a reduced capture efficiency. Additionally, the intra-annual variation in temperature would have a nonlinear effect on evaporation, and hence reservoir drawdown, that is not depicted by a long-term change in annual value.

The second major assumption is the aggregation of hydrometeorology and water uses over four large hydroclimatological regions of the state. The reservoirs in each region are taken to be lumped together, in terms of area and storage, and placed at the dam site farthest downstream in the region. This maximizes the efficiency of that hypothetical reservoir, capturing the greatest amount of runoff feasible for its size. The reality is that reservoirs are distributed throughout the basin, each with a limited effectiveness in the amount of runoff it can intercept and store. Moreover, the actual runoff in a basin is skewed to the east, so that a higher proportion is unavailable for human storage and use. The water budget is carried out for transfers only and does not consider the volume of water in each compartment. Notably, storage and drawdown of reservoirs are not modeled; spills are therefore the net of runoff less evaporative losses and withdrawals. Thus the estimate of the ability of the reservoirs to meet the demands is optimistic.

The third major assumption underlies the central feature of the water-budget model presented here, the set of runoff-versus-precipitation regressions developed by the USGS (Lanning-Rush 2000). These equations capture, in part, the decrease in the ratio of runoff to rainfall with decreasing rainfall, but the use of annual runoff based on annual precipitation means that much

TABLE 3.8. Central region water budget components for various scenarios

WATER COMPONENT[a]	SCENARIO			
	NORMAL CLIMATE	GREENHOUSE-WARMED	DROUGHT NORMAL	GREENHOUSE DROUGHT
Precipitation	100	95	80	76
Evapotranspiration	100	96	85	81
Runoff	100	81	42	34
Recharge	100	95	80	76
Lake evaporation	100	120	110	132

HUMAN WATER USE	WATER-USE SCENARIO YEAR						
	2000	2050	2000	2050	2000	2050	2050
Surface water							
municipal-industrial	100	193	106	205	102	195	207
agriculture	100	132	134	176	110	145	189
electric	100	274	130	356	107	294	368
Groundwater							
municipal-industrial	100	178	106	189	101	179	190
agriculture	100	119	134	159	110	131	171
electric	100	278	130	360	107	297	373
Return flows							
municipal-industrial	100	188	106	199	101	190	201
agriculture	100	126	134	168	110	138	181
Steam-electric circulation	100	274	104	286	101	278	288
Downstream flow to							
Texas coast	100	94	71	64	24	22	26

[a] Measured as a fraction (percent) of Present Normal conditions

TABLE 3.9. South region water budget components for various scenarios

WATER COMPONENT[a]	SCENARIO			
	NORMAL CLIMATE	GREENHOUSE-WARMED	DROUGHT NORMAL	GREENHOUSE DROUGHT
Precipitation	100	95	71	67
Evapotranspiration	100	95	71	68
Runoff	100	82	26	21
Recharge	100	95	71	67
Lake evaporation	100	116	119	137

HUMAN WATER USE	WATER-USE SCENARIO YEAR						
	2000	2050	2000	2050	2000	2050	2050
Surface water							
municipal-industrial	100	205	112	229	103	211	235
agriculture	100	130	134	174	110	143	187
electric	100	600	126	755	107	643	794
Groundwater							
municipal-industrial	100	154	112	172	103	159	177
agriculture	100	88	134	117	110	97	126
electric	100	277	126	349	107	297	367
Return flows							
municipal-industrial	100	187	112	209	103	192	214
agriculture	100	112	134	150	110	123	161
Downstream flow to							
Texas coast	100	80	67	41	11	18	20

[a] Measured as a fraction (percent) of Present Normal conditions

TABLE 3.10. Reservoir surfeit flows for hydroclimatological regions of Texas, under normal and drought models

REGION[a]	SCENARIO AND WATER-USE YEAR[b]						
	NORMAL CLIMATE		GREENHOUSE-WARMED NORMAL		DROUGHT		GREENHOUSE DROUGHT
	2000	2050	2000	2050	2000	2050	2050
High Plains	0.0	−0.1	−0.1	−0.2	−0.2	−0.3	−0.4
Central	17.1	14.0	11.4	8.0	2.4	−0.7	−3.9
East	15.1	14.5	12.1	11.3	5.5	4.8	3.0
South	2.3	1.5	1.4	0.5	−1.6	−2.4	−3.2
State	34.4	29.9	24.7	19.5	6.1	1.4	−4.4

[a] As defined in Figure 3.7

[b] Water surfeit in millions of acre-feet per year. Negative numbers indicate a shortfall.

of the effect of correlated high flows with wet conditions and low flows with dry is averaged out of the data before the regressions are determined. Much more runoff (per unit of rainfall) occurs when the watershed is saturated during the wet season than is reflected in the equations, and likewise much less runoff occurs when the watershed is desiccated during the dry season. It is this intra-annual variation that is expected to be particularly exacerbated

TABLE 3.11. Flows to the Texas coast, for the Central region and statewide, under normal and drought models

REGION[a]	SCENARIO AND WATER-USE YEAR[b]						
	NORMAL CLIMATE		GREENHOUSE-WARMED NORMAL		DROUGHT		GREENHOUSE DROUGHT
	2000	2050	2000	2050	2000	2050	2050
	FLOWS IN MILLIONS OF ACRE-FEET PER YEAR						
Central	17.0	16.0	12.1	10.9	4.2	3.1	0.5
East	17.7	17.4	14.5	14.0	7.1	6.8	5.0
South	2.7	2.2	1.8	1.1	0.0	0.0	0.0
State	37.5	35.6	28.4	26.1	11.3	9.9	5.5
	FLOWS AS A FRACTION (PERCENT) OF 2000 NORMAL						
Central	100	94	71	64	24	18	3
East	100	98	81	79	40	38	28
South	100	80	67	41	0	0	0
State	100	95	76	69	30	26	15

[a] As defined in Figure 3.7

[b] Assumes that the deficits in Table 3.10 are met with new supplies in the East and Central regions

by climate change. This will also have a strong geographical variation within the four hydroclimatological regions used here. None of this is depicted in the water-budget model; instead, the extremes of runoff as a function of rainfall are diminished and the climate-change response muted.

The fourth assumption is the set of specific causal pathways by which climate change affects the water budget represented in this model: (1) runoff reductions due to decreased rainfall; (2) increased evaporation, forced evaporation (power generation), and increased municipal and industrial demands as a function of increasing temperature; (3) increased agriculture demand with increasing temperature. There are other relations driven by the two variables of rainfall and temperature, for example, reduced precipitation forcing increased demands and increased temperature driving increased evapotranspiration, hence reduced runoff. There are other climate-change variables not considered, such as decreased clouds and increased carbon dioxide concentration, most of which will further augment the changed-climate response.

Some of these weaknesses can be repaired by the more complex procedure of water-budget analysis on a regional scale, with more refined time resolution. Schmandt and Ward (1991) reported such an analysis for three river basins in Texas (the Trinity, Colorado, and Rio Grande), and later Schmandt et al. (2000) reported a more detailed evaluation of the Rio Grande. For all of these basins, severe reservoir drawdowns were determined; in most cases, the storage was depleted before the end of the drought of record. Wurbs et al. (2005) report an analysis of the Brazos basin achieved by coupling a detailed watershed model (SWAT) to the state water availability model, thereby allowing monthly simulations and detailed reservoir operation. The results are generally consistent with those presented here (though the predicted changed-climate scenarios were based on only a single global climate model, and the relations between atmospheric parameters and watershed response are not clear).

These data demonstrate the extent to which Texas is vulnerable to changes in climate. The drought of the 1950s is within living memory, and yet it is evident that population growth alone would make it extremely difficult to cope with a similar drought under the 2050 scenario, during which many water uses would have to be curtailed. When the consequences of global warming for Texas climate are included in the analysis, the situation is even more serious, a conclusion that is even more robust in light of the minimal responses assumed for the water-budget components in this analysis.

REFERENCES

Fehrenbach, T. R., 1983. *Seven Keys to Texas*. Texas Western Press, El Paso.

IPCC, 2007. *Climate Change 2007: The Physical Science Basis*. Contribution of Working Group 1 to the Fourth Assessment Report of the Intergovernmental Panel on Climate Change. Cambridge University Press, Cambridge, U.K., and New York.

King, C., I. Duncan, and M. Webber, 2008. Water Demand Projections for Power Generation in Texas. Report to Texas Water Development Board (Contract 0704830756), Bureau of Economic Geology, University of Texas, Austin.

Lanning-Rush, J., 2000. Regional Equations for Estimating Mean Annual and Mean Seasonal Runoff for Natural Basins in Texas, Base Period 1961–90. Water-Resources Investigations Report 00-4064. U.S. Geological Survey, Austin, Tex.

Mitchell, J. F. B., 1989. The Greenhouse Effect and Climate Change. *Reviews of Geophysics* 27(1):115–139.

Muller, D., and R. Price, 1979. Ground-water Availability in Texas. Report 238. Texas Department of Water Resources, Austin.

Schmandt, J., and G. Ward, 1991. Texas and Global Warming: Water Supply and Demand in Four Hydrological Regions. Policy Research Project Report, LBJ School of Public Affairs, University of Texas, Austin.

Schmandt, J., I. Aguilar-Barajas, M. Mathis, N. Armstrong, L. Chapa-Alemán, S. Contreras-Balderas, R. Edwards, M. García-Ramirez, J. Hazleton, M. Lozano-Vilano, J. Navar-Chaidez, E. Vogel, and G. Ward, 2000. Water and Sustainable Development in the Binational Lower Rio Grande/Río Bravo Basin. Final Report to EPA/NSF Water and Watersheds grant program (Grant No. R 824799-01-0). Houston Advanced Research Center, Center for Global Studies, The Woodlands.

Texas Water Development Board, 2007. Water for Texas 2007, Vols. 1-3. Doc. GP-8-1. Texas Water Development Board, Austin.

Ward, G. H., 1993. A Water Budget for the State of Texas with Climatological Forcing. *Texas Journal of Science* 45(3):249–264.

Ward, G. H., 2004. Texas Water at the Century's Turn: Perspectives, Reflections and a Comfort Bag. In: *Water for Texas*. J. Norwine, J. Giardino, and S. Krishnamurthy (eds.). Texas A&M University Press, College Station.

Wurbs, R., R. Muttiah, and F. Felden, 2005. Incorporation of Climate Change in Water Availability Modeling. *Journal of Hydrological Engineering* 10(5):375–385.

CHAPTER 4

Coastal Impacts

Paul A. Montagna, Jorge Brenner, James Gibeaut,
and Sally Morehead

The Texas coast is likely to experience severe climate change impacts because of a synergy between the regional climate regime and the coastal geology. Lying between about 26 and 30 degrees N latitude, the Texas coast is already in a relatively warm climate zone and subject to very high rates of evaporation (Larkin and Bomar 1983). Thus, potential changes in rainfall or temperature will have great impacts on the Texas coastal hydrocycle. The Texas coastal plain is relatively flat and low-lying, and the coast also has one of the highest rates of subsidence in the world (Anderson 2007). Thus, changes in sea level will be exacerbated on the Texas coast because the land is relatively flat and it is rapidly sinking. The combined effects of these changes can affect the physical and biological characteristics of the Texas coast dramatically.

In one of the earliest discussions of the potential impacts of climate change along the Texas coast, Longley (1995) focused on potential changes in habitat area that might result from changes in precipitation and concomitant changes in freshwater inflow to bays and estuaries. Other authors have focused on sea-level rise (Zimmerman et al. 1991) or temperature change (Applebaum et al. 2005). In addition, Twilley et al. (2001) provided a comprehensive assessment of climate drivers, such as changes in temperature, rainfall, freshwater resources, and sea-level rise, and the consequences of human activities as they act in concert with climate change effects.

If the Texas coast is indeed exceptionally susceptible to climate change effects, then there must be both physical and biological indicators of change. Temperature change itself is an obvious indicator. Salinity is an indicator of changes in the freshwater cycle, because fresh water dilutes sea water when it flows to the coast. It is also possible for indirect changes of water quality to occur because oxygen is less soluble in hotter, saltier water. Thus, the temporal dynamics of changes in water quality are also an indicator. Species that are sensitive to changes in any one or more of these factors, or reside at the edge of their distribution range, are indicator species.

In the context of climate change, the indicator species are sensitive to temperature, salinity, or elevation changes. One potential indicator species is the black mangrove (*Avicennia germinans*), because its distribution and survival in Texas are limited by winter temperature (Sherrod and McMillan 1981). Other indirect effects include explicit links between temperature and water quality and change in biotic responses. The earlier habitat change analysis conducted by Longley (1995) assumed that only inflow rates will change, but rising sea levels may obliterate these effects; thus attention to effects of sea-level rise are critical. In the current study, the focus is on identifying changes in the instrumental record (for both water and habitats) to determine if there are trends in recent long-term records of water temperature and quality, mangrove habitat cover, and sea-level rise.

TEXAS ESTUARIES

The Texas Gulf shoreline stretches 370 miles from the Sabine River at the Louisiana state line to the Rio Grande at the Mexican border. Except for two areas along the upper coast, narrow barrier islands and peninsulas separate the Gulf of Mexico from the shallow estuaries. Pritchard (1967) defines *estuary* as a semi-enclosed coastal body of water that has a free connection with the open sea and within which sea water is measurably diluted with fresh water derived from land drainage. The estuaries behind the barrier islands and peninsulas project inland from the Gulf shoreline as much as 30 miles. The land surrounding these aquatic systems is low and flat; one must travel 30–50 miles inland from the Gulf shoreline to reach a land elevation of just 100 feet above mean sea level. The environments of the coastal region can be looked at from several scales of view: the region as a whole, individual estuaries, and habitats. Each viewpoint is useful in understanding the effects of climate and climate change on estuaries.

River Basins

The land areas from which runoff drains to rivers can be delineated on topographic maps through careful consideration of land elevation and slope. One or more river basins may drain to an estuary on the coast. The amount of fresh water that flows to an estuary has a strong influence on the shape and form of the estuary, as well as on the habitats and organisms there and in the surrounding wetlands. From a regional viewpoint, the river basins of the entire state drain to 11 estuaries (Fig. 4.1).

Drainage basins for rivers that flow into the Brazos, Colorado, and Rio Grande estuaries extend across the entire state. Flows of water to these estuaries are influenced by rainfall and runoff from a large area, much of which is far removed from the coastline. The basin for the Rio Grande even extends into Mexico, New Mexico, and southern Colorado. Thus, climate changes occurring a long distance from Texas estuaries may nevertheless have a profound influence on the amount of fresh water that flows into them.

The Sabine-Neches, Trinity–San Jacinto, Guadalupe, and Nueces estuaries have smaller river basins that extend only partway across the state. The Sabine-Neches Estuary also receives some runoff from Louisiana. These smaller river basins are not directly affected by precipitation in the western half of the state. The San Bernard, Lavaca–Tres Palacios, Mission-Aransas, and Laguna Madre estuaries have the shortest river basins; they reach inland no more than about 100 miles from the Gulf shoreline. Runoff to these estuaries is most strongly influenced by the climatic conditions close to the coast.

The runoff rates vary by more than an order of magnitude between the basins bordering Louisiana and those bordering Mexico. Differences in runoff rates are the result of the interaction of the east-west precipitation and evaporation gradients in the state (described in Chapter 2), the location of the basin along the coastline, and the distance the drainage basin extends inland from the Gulf shoreline. In general, the more a basin is restricted to the area near the coast, or the farther to the northeast it is located, the greater the average runoff per square mile. Basins that extend farther across the state or are located more to the southwest have lower rates of runoff.

Estuary Forms and Climate

The development and configuration of estuaries on the Texas coast are closely bound to climatic changes that have taken place over the period from the last ice age to the present. Three major shapes or forms of estuaries can be seen on the Texas coast (Fig. 4.1). The first is the classic estuarine configuration, such as the Trinity–San Jacinto Estuary, which juts far inland and has an opening to the sea toward its southern tip. Rivers discharge at the most inland end of the estuary, and the axis between the river mouth and the openings to the sea allows the establishment of a long gradient of salinity ranging from low to high levels. Most of the land that surrounds this estuary and others with the same physical form (Sabine-Neches, Lavaca–Tres Palacios, Guadalupe, Mission-Aransas, and Nueces) is of Pleistocene age (Fisher et al. 1972) and was formed 60,000–1 million years before the present (B.P.). During the last ice age (Late Wisconsin glaciation), which began about 50,000–60,000 years B.P., sea level declined as water was captured in the ice sheets that covered part of the Northern Hemisphere. As sea level fell to its lowest point, 300–400 feet below its present level, the

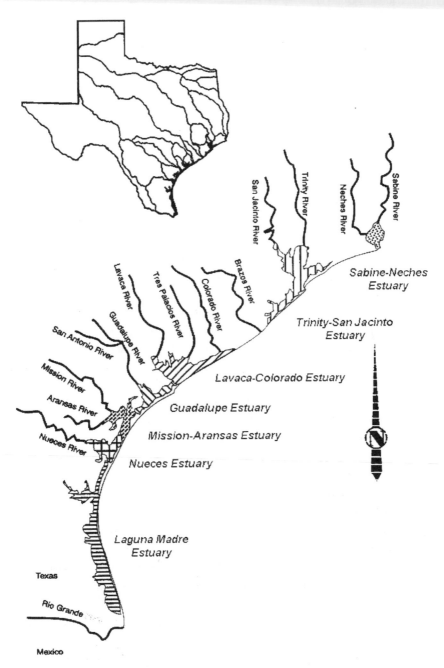

FIGURE 4.1. Texas estuaries (Longley 1994). Inset: Extension of major river basins across the state.

land where today's estuaries and near-shore Gulf are located was completely exposed; surface sediment eroded, creating deep river valleys. Rivers flowed through these valleys at levels 100–130 feet lower than today's mean sea level. As the climate began to warm, about 20,000 years B.P., sea level started to rise and began to flood the river valleys. The deeper portions of the valleys gradually filled with sediment originating in upland runoff, offshore currents, and eroding valley walls. Offshore sediments that were transported onshore also contributed to the formation of barrier islands and peninsulas, which formed within the past 2,500 years. The processes of transport, deposition, and erosion have continued to the present, slowly filling the deeper areas of the estuaries and widening the edges of the Sabine-Neches, Trinity–San Jacinto, Lavaca–Tres Palacios, Guadalupe, Mission-Aransas, and Nueces estuaries.

River basins that drain to these classic-form estuaries are generally midsized or coastal basins and do not extend more than halfway across the state. The existence and shape of these estuaries is clearly a result of the climatic shift that occurred during the ice ages, the subsequent rise of sea level, and the transport and erosion of sediment affected by varying climatic conditions throughout the river basin area.

A second estuarine configuration is illustrated by the Brazos, Colorado, and Rio Grande river mouths. The Brazos and the Rio Grande now empty directly into the Gulf rather than into a bay. The Colorado River has recently been artificially diverted into the eastern portion of Matagorda Bay and no longer flows directly to the Gulf. During the last ice age, these rivers cut deep valleys just like rivers flowing into the classic estuaries. As the climate warmed and sea level rose, however, these rivers carried so much sediment that their deltas expanded, completely filling their river valleys. This deposition continued all the way to today's Gulf shoreline, allowing the rivers to discharge directly into the sea. The great sediment load of these three rivers resulted from the large areas of their drainage basins, which extend across the entire state. The land immediately surrounding these estuaries was formed very recently. The rapid land growth and direct discharge into the Gulf of the river estuaries is the result of the interaction of climate, physiography, and soil type in the river basins over the past 5,000–10,000 years.

Initially, the San Bernard River Estuary, just south of the Brazos River, seems to be an exception to the generalization that the form of the river estuaries is the result of sedimentation in the large river basins that stretch across the state. Today the San Bernard has a very small basin, located between the present Brazos and Colorado river basins. The San Bernard River is most likely an abandoned channel of one or the other of these major rivers. When sea level approached its present level, about 2800 B.P., the Colorado and Brazos rivers emptied into a common estuary and filled it in about 1,200 years (McGowen et al. 1976). Because the present San Bernard basin is entirely within the area affected by the Colorado and Brazos sedimentation, this estuary and the land surrounding it were only recently formed by the large sediment loads of these two great river basins.

The third form of estuarine system on the Texas coast is the lagoon, typified by the Laguna Madre. The water body is narrow, and the major axis runs parallel to the shoreline rather than perpendicular to it. Except for the area close to the Rio Grande, the land on the inland side of the laguna was formed as fluvial-deltaic deposits or strandplain during the Pleistocene (Brown et al. 1977, 1980). The barrier island side of the laguna was formed over the past 2,500 years, as sea level rose to its present position. Thus, even the current configuration of the Laguna Madre (actually two different systems, the Upper Laguna Madre–Baffin Bay and Lower Laguna Madre) is the result of a rise in sea level as the climate returned to its present, interglacial state.

Changes in Shoreline

Most of the sandy Gulf of Mexico shoreline of South Texas has probably been retreating for several thousand years, definitely since the mid to late 1800s, when sufficiently accurate shoreline maps were constructed for comparison with today's maps. An analysis of multiple Gulf of Mexico shorelines from the 1930–2000 time period and from the Colorado River to the Rio Grande shows that 56 percent of the shoreline retreated at a mean rate of 7.2 feet per year, 36 percent was essentially stable, and only 8 percent advanced seaward. The advancing shoreline sections were associated with impoundment of sand by jetties or spit progradation caused by engineering alterations affecting Pass Cavallo. A section a few miles long in the central Padre Island area also advanced because of the natural convergence of littoral drift.

Bay shorelines have been retreating for at least 10,000 years as sea level rose from the lowstand of 18,000 years ago and flooded paleo river channels running through the bays. Inundation, waves, and tidal action eroded the riverbanks, and the resulting shoreline retreat largely shaped the bays as they exist today. Generally, these bay shorelines continue to retreat with the erosion of marshes and flats, clay bluffs, sandy slopes, and sand and shell beaches. In some areas, extensive shore protection structures such as riprap and bulkheads have been installed. Paine and Morton (1993) determined an average retreat rate for the Copano, Aransas, and Redfish bay systems of 0.8 feet per year from 1930 to 1982. In Baffin Bay, most of the shoreline retreated from 1941 to 1995 (Gibeaut and Tremblay 2003).

EFFECTS OF CLIMATE CHANGE ON COASTAL REGIONS
Sea-Level Change

Changing sea level relative to the land (relative sea-level change) and the increase and decrease in sand supply to the coast cause shorelines to retreat or advance over a period of 100 years or more (Bruun 1962; Gibeaut and Tremblay 2003). The rise in relative sea level during the last 100 years along the Texas coast has moved the Gulf and bay shorelines through inundation and by shifting the erosive energy of waves and currents landward. This has happened because, overall, the rate of new sediment delivered to the littoral zone has not been sufficient to stem the effects of relative sea-level rise. Localized exceptions to this are where rivers form deltas at the heads of the bays, such as the Nueces and Mission deltas, where creeks erode bluffs and enter the bays (Morton and Paine 1984; Paine and Morton 1993), and where dunes have migrated and advanced the shoreline (Gibeaut and Tremblay 2003; Prouty and Prouty 1989). Because of this sediment deficit and the low-lying and gently sloping shores of much of the South Texas coast, relative sea-level rise has had and will continue to have a profound effect on coastal habitats. Increases in the rate of global sea-level rise, as projected by global climate modeling (IPCC 2007), and coastal development will very likely result in further decreases of coastal wetland habitats.

Relative sea-level rise along the South Texas coast is caused by natural and human-induced land surface subsidence and a global rise in ocean level. Global sea level is rising primarily through the addition of water to the oceans by melting continental ice and, to a lesser degree, through thermal expansion of ocean water (Miller and Douglas 2004), both of which are caused by global warming. Tide gauge records in South Texas, which include the effects of land subsidence, show that relative sea level has risen at a rate of 4.6 millimeters (0.18 inches) per year at Rockport since 1948, 2.05 millimeters (0.08 inches) per year at Port Mansfield since 1963, and 3.44 millimeters (0.14 inches) per year at South Padre Island since 1958 (Zervas 2001). Douglas (1991) considered tide gauges from around the world and, after accounting for vertical land movements, determined that global sea level from the late nineteenth to the late twentieth century rose at a rate of only 1.8 millimeters (0.07 inches) per year. Land subsidence rates can be estimated for the tide gauge locations by subtracting 1.8 millimeters (0.07 inches) per year from the relative sea-level rise rate recorded by the gauge. This illustrates that land subsidence is an important component of relative sea-level rise along the Texas coast.

Additional land subsidence is caused by groundwater withdrawal and oil and gas production, which decreases pore pressures in underlying sediments, allowing further compaction. Ratzlaff (1980) compared releveling surveys for various periods from 1917 to 1975 and observed locally high land subsidence rates of as much as 49 millimeters (1.93 inches) per year, such as at the Saxet oil and gas field southwest of Nueces Bay. The highest rates correlated with oil and gas and groundwater production. By combining tide gauge and releveling data, Paine (1993) estimated Texas coastal subsidence rates of 3–7 millimeters (0.12–0.28 inches) per year. Sharp et al. (1991) and Paine (1993) hypothesized that regional depressurization of petroleum reservoirs was the reason historical subsidence rates were much higher than the geologically long-term rates. Morton et al. (2006) provided evidence of hydrocarbon production causing regional land subsidence and associated wetland loss in the Mississippi Delta and the upper Texas coast regions.

The IPCC (2007) report provides model projections for global sea-level rise based on six greenhouse gas and aerosol emission scenarios. The range in the amount of projected global sea-level rise by 2099 relative to the average from 1980 to 1999 is 0.18–0.59 meters (0.6–1.9 feet) per year. After estimates for local land subsidence are added, the amount of projected relative sea-level rise by the year 2100 is 0.46–0.87 meters (1.5–3 feet) at Rockport, 0.2–0.61 meters (0.66–2 feet) at Port Mansfield, and 0.34–0.75 meters (1.1–2.5 feet) at South Padre Island. These amounts will likely be greater in areas with relatively thick Holocene deposits filling paleo river channels and tidal inlets, such as along the barrier islands and modern deltas at the heads of the bays (e.g., Nueces River Delta), and they may be much higher where subsidence caused by groundwater and hydrocarbon extraction occurs.

Depositional subenvironments of barrier islands and bay margins are the substrates for various types of aquatic, wetland, and upland habitats. These subenvironments and associated habitats are closely linked to elevation relative to sea level through the processes that form and maintain them. On the low-lying, sandy barrier islands of the microtidal Texas coast (tide range 0.6 meter, or 2 feet, on the open coast and less than 0.3 meter, or 1 foot, in the bays), a rise of just 0.1 meter (0.3 feet) in relative sea level can cause conversion of fringing low marshes and flats to open water and seagrass beds, and of usually dry high marshes and flats to usually wet low marshes and flats (Gibeaut et al. 2003).

Mustang Island is a barrier island at the mouth of Corpus Christi Bay (Fig. 4.2). Most of the island, except for the tallest foredunes, is less than 3 meters (10 feet) above sea level, which is typical for the barrier islands along the south Texas coast. The inundation of Mustang Island has been projected for 2100, at various amounts of relative sea-level rise (Fig. 4.3). A rise of just 0.45 meter (1.5 feet) will cause lateral shifts of 1–2 kilometers (0.6–1.2 miles) of the bayside wetland environments. The upper limit of the projection (a rise of 3 feet) will narrow the upland areas of the island to a width less than 200 meters (656 feet) in places and flood central portions of the city of Port Aransas on the north end of the island.

Besides the rise in relative sea level, actual shoreline retreat and loss of wetland areas on Texas barrier islands will depend on (1) vertical sediment accretion and its ability to keep up with the rise, (2) the slope of adjacent uplands and whether it is gentle enough to allow wetlands to migrate landward, (3) any obstruction that development presents to the upward or landward migration of wetlands, and (4) the severity of the erosion at the edge of marshes and flats by waves and currents. Based on the observed conversion

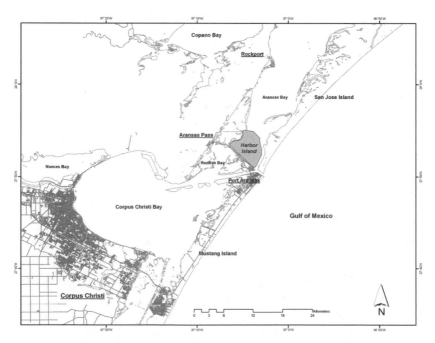

FIGURE 4.2. Nueces Estuary on the Texas coast.

of tidal flats to open water and seagrass beds and the migration of marshes into higher areas since the 1950s on Mustang and adjacent islands (White et al. 2006), it is unlikely that vertical accretion will offset the effects of an increase in the rate of relative sea-level rise. Upland slopes increase toward the core of the islands, which will temper the amount of new marsh that can develop, and it is very likely that future development will obstruct new marsh creation, such as occurred at the Padre Isles development (Fig. 4.3) beginning in the 1970s (White et al. 2006). Erosion of the outer edges of

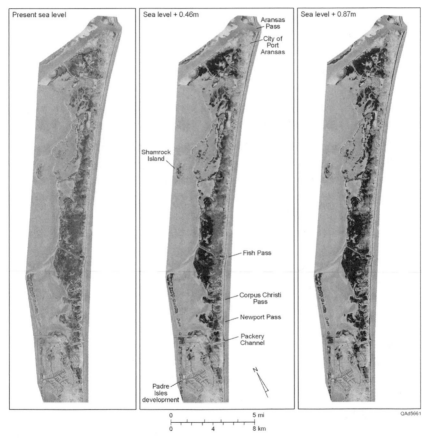

FIGURE 4.3. Inundation of Mustang Island. In 100 years, sea level (lightest gray) is expected to rise between 1.5 feet (center panel) and 3 feet (right panel) above its present level (left panel). These results combine local subsidence estimates with the lower and upper ranges of global sea-level rise projections (IPCC 2007). The maps were created using aerial photography draped on a high-resolution, lidar-derived digital elevation model. Lidar data acquisition and processing were performed by the Bureau of Economic Geology, University of Texas at Austin, in 2005.

marshes and flats since the 1930s has caused shoreline retreat in the range of 0.5–2.5 meters (1.6–8 feet) per year along most of the Mustang Island bay shoreline (Morton and Paine 1984; Williams 1999). Erosion by waves may increase as higher water decreases the amount of wave shoaling and extends wave energy farther landward.

As sea level rises and the barrier islands become narrower, large storms will eventually breach, wash over, and transport sand landward into the bays. This process is already happening along low and narrow portions of the South Texas coast, as in the Corpus Christi and Newport passes (Fig. 4.3) during Hurricane Beulah in 1967 (Davis et al. 1973). Furthermore, increasing aridity, which climate models predict for this region, has the potential to reduce stabilizing dune vegetation and cause more active dune migration and blowouts, as observed on North Padre and Mustang islands during the drought of the 1950s (Prouty and Prouty 1989; White et al. 1978). Hurricane intensities have increased recently and may continue to increase with warming sea surface temperatures (Emanuel 2005). Hence, rising sea level, increasing aridity, and increasing storm intensity will drive the Texas barrier islands toward narrower, lower-lying islands that are more frequently washed over and severed by tropical storms. Eventually, depending on the actual rate of relative sea-level rise, portions of the Texas barrier island chain will be destroyed, similar to the near demise of the Chandeleur Islands in Louisiana following recent hurricanes.

The IPCC (2007) climate projections for global sea-level rise do not include the full effects of potential changes in polar ice sheet flow as global warming proceeds. Recent observations of ice sheet changes suggest the possibility of large contributions to sea-level rise from the flow of Greenland and Antarctic glaciers into the oceans (Rignot and Kanagaratnam 2006; Shepherd and Wingham 2007). Overpeck et al. (2006) compared 100-year global temperature projections in the IPCC Third Assessment Report (IPCC 2001) with climatic and sea-level conditions during the last interglacial period, about 130,000 years ago. During that time, polar temperatures were 3–5°C (5.4–9°F) higher than now, causing polar ice to retreat and contributing to a sea level that was 4–6 meters (13–20 feet) higher than today. This amount of warming is within the range projected during the next 100 years by IPCC modeling studies. Increases in sea level caused by melting polar ice, therefore, may be expected to proceed for centuries, resulting in a sea-level rise of several meters and rise rates twice those projected in the IPCC (2007) report (Overpeck et al. 2006).

Applying that scenario to the Texas coast, with sea level rising 6 meters (20 feet), the barrier islands are completely inundated, and the sea advances about 20 kilometers (12.5 miles) inland from the bay margins, except where sufficiently high Pleistocene bluffs and uplands exist (Fig. 4.4). River valleys and deltas at the heads of the secondary bays are flooded. White et al. (2002) showed that vertical accretion rates on the Nueces delta are less than the rate of relative sea-level rise today. It is unlikely that vertical accretion rates will be sufficient to maintain wetlands on the South Texas deltas when relative sea-level rise rates are 10 millimeters (0.4 inches) per year or more, as in the Figure 4.4 scenario. If polar ice sheet destabilization should occur, therefore, we can expect massive losses of critical wetland habitat in Texas bays.

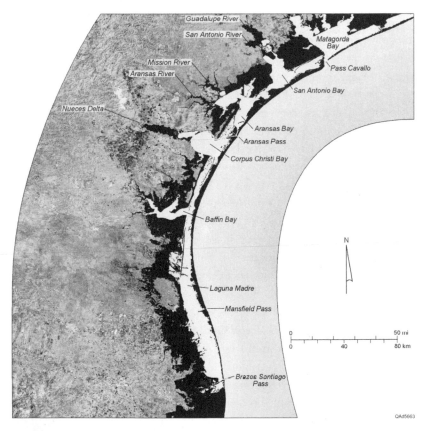

FIGURE 4.4.   Inundation of the South Texas coast if the level of the sea were to rise 20 feet above the present level. Lighter gray shows the present bays and lagoons; darkest gray represents inundated areas. This scenario assumes melting of the polar ice sheet and destabilization triggered by global warming during the next 100 years. Constant redistribution of sediments by waves and currents during the rise in sea level would tend to smooth the shoreline, but those processes are not reflected here.

Water Quality

Water quality data were obtained from the website of the Texas Water Data Services (2008), and they originate from the Coastal Fisheries Division of the Texas Parks and Wildlife Department (TPWD). The TPWD samples all Texas bays monthly using a probabilistic sampling design (i.e., stations within bays are randomly sampled). While collecting organisms at trawl stations in the bays, the TPWD simultaneously collects water data using multiparameter sondes, measuring salinity, temperature, and dissolved oxygen at 1 foot from the bottom.

Texas follows the traditional system of naming an estuary for the river that dilutes the sea water (Longley 1994). In publications of the National Oceanic and Atmospheric Administration (e.g., Orlando et al. 1993), these systems are named after the primary bay (Sabine Lake, Galveston Bay, Matagorda Bay, San Antonio Bay, Aransas Bay, Corpus Christi Bay, and Laguna Madre). In addition, TPWD samples in the Gulf of Mexico, making a total of eight marine systems measured.

Under this system, TPWD recorded a total of 173,735 water quality observations between 1976 and 2007. Because the number of samples varies among bay-month combinations, the first step in data reduction was to calculate means by bay-month cell. Then the data were averaged by year to analyze for year-year differences among bays and the Gulf of Mexico.

An analysis of covariance (ANCOVA) was used to analyze the water quality data, because they were reduced to one annual sample taken in each region. Thus, regions are blocks, and the year is the covariate. The year-bay system interaction was used to test for common responses among regions (called a test for heteroscedasticity). All data management, data aggregation, reduction, and ANCOVA were conducted using SAS 9.1 software (SAS Institute 2004, 2006).

The climatic gradient along the Texas coast influences freshwater inflow to estuaries. This gradient of decreasing rainfall and concomitant freshwater inflow, from northeast to southwest, is the most distinctive feature of the coastline. Along this gradient, rainfall decreases by a factor of 2, from 142 centimeters (56 inches) per year near the Louisiana border to 69 centimeters (27 inches) per year near the border with Mexico (Larkin and Bomar 1983). Inflow balance, however, decreases by almost two orders of magnitude, from about 17 billion cubic meters per year in Sabine Lake to about 900 billion cubic meters per year in Laguna Madre (Longley 1994). Inflow balance is the sum of freshwater inputs (gauged, modeled runoff, direct precipitation, plus return flows) minus the outputs (diversions and evaporation). The net effect is an increase in salinity as we move down the Texas coast from the

Louisiana border, from an average salinity in Sabine Lake of 7 parts per thousand (ppt) to 36 ppt in Upper Laguna Madre (Table 4.1), a hypersaline lagoon (the nearshore Gulf of Mexico averages 31 ppt).

There has been a long-term trend of increasing water temperature along the entire coast (Fig. 4.5). The patterns over time differ among the estuary systems. The main difference is a higher rate of increase in Lower and Upper Laguna Madre than in the other seven ecosystems. The overall average rate of increase in temperature is 0.0428°C per year, which translates into an increase of 1°C in 23 years (1°F in 13 years).

TABLE 4.1. Water quality parameters for Texas bay systems, 1976–2007

BAY SYSTEM	ANNUAL MEAN OF 32 YEARS (STANDARD DEVIATION)							
	SALINITY (PPT)		TEMPERATURE (°C)		DISSOLVE OXYGEN (MG/L)		TURBIDITY (NTU)	
Gulf of Mexico	30.52	(1.16)	22.19	(0.55)	7.09	(0.56)	11.88	(5.13)
Sabine	7.07	(2.83)	21.92	(1.05)	8.08	(1.02)	25.93	(6.03)
Matagorda	19.98	(4.58)	22.67	(0.75)	7.98	(1.03)	35.12	(13.49)
Galveston	16.03	(3.18)	22.23	(0.68)	8.28	(0.98)	38.53	(24.52)
San Antonio	16.81	(4.90)	22.82	(0.64)	7.99	(0.87)	29.40	(10.77)
Aransas	18.40	(5.59)	22.93	(0.78)	8.25	(0.73)	30.86	(16.74)
Corpus Christi	28.99	(3.78)	23.22	(0.81)	7.59	(0.72)	29.77	(24.44)
Upper Laguna Madre	35.94	(7.00)	24.07	(1.00)	7.22	(0.67)	36.08	(25.63)
Lower Laguna Madre	32.07	(2.74)	24.36	(1.01)	7.82	(0.71)	35.05	(23.92)

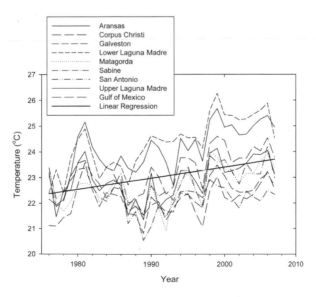

FIGURE 4.5. Average annual water temperature in Texas coastal ecosystems.

In contrast, dissolved oxygen has decreased in these coastal ecosystems over the long term (Fig. 4.6). Again, the patterns differ among the estuaries. The main difference is a higher rate of decrease in Galveston Bay and Upper Laguna Madre than in the other seven ecosystems. The overall average rate of decrease in dissolved oxygen is approximately 0.7 percent per year.

Salinity patterns are similar among all the ecosystems, with a probability (P) of 0.6470, but there are statistically significant differences in salinity among the regions (P = 0.0001, Table 4.1). These differences are related to the climatic gradient from east to west along the Texas coast. There is no evidence of salinity change over time (P = 0.5464). The increase in temperature and decrease in dissolved oxygen was noted previously in just Corpus Christi Bay (Applebaum et al. 2005). Both rates for the period 1982–2002 were greater than reported here for 1976–2007. The initial report was that temperature increased at a rate of 0.07°C per year and dissolved oxygen decreased at a rate of 0.06 milligrams per liter per year. The occurrence of

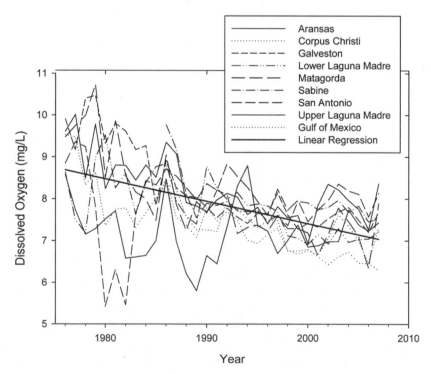

FIGURE 4.6. Average annual dissolved oxygen concentrations in Texas coastal ecosystems.

this joint observation everywhere in Texas indicates the general nature of the indirect effects of air temperature increases. The relationship between temperature and dissolved oxygen is causal because oxygen solubility in water decreases with increasing temperature. It appears that increased sea temperatures will lead to decreased dissolved oxygen and water quality in many places around the world. This may, in part, explain the increasing number of hypoxic zones worldwide (Diaz and Rosenberg 2008).

As found earlier by Applebaum et al. (2005), there was no trend in salinity with time in the current coastwide analysis. The importance of global-scale climate events in driving regional-scale climatic variability and salinity structure in Texas estuaries is now reasonably well understood (Tolan 2007). The El Niño–Southern Oscillation (ENSO) index is correlated to salinity structure within Texas estuaries within 4–6 months. During ENSO events, salinities in Texas estuaries decrease because of increased rainfall and freshwater flows to the coasts. During La Niña periods, salinities increase because of the drier climatic conditions. These cycles occur with a periodicity of 3.55, 5.33, and 10.67 years. The ENSO is dominated by the 3.55- and 5.33-year periods, and the Pacific Decadal Oscillation is defined by the 10.67-year period. The link with ENSO explains the year-to-year variability in salinity along the Texas coast.

Other evidence indicates that there are long-term trends of lower rainfall in the Gulf of Mexico and the American Southwest. With increased temperatures and decreased rainfall, a water deficit would be expected where evaporation exceeded rainfall. These are the same conditions that cause hypersalinity in Laguna Madre. Between 1987 and 2006, the Gulf of Mexico had a strong water deficit, as did most of the Pacific Ocean between about 15 and 30 degrees N latitude (Wentz et al. 2007). Between about 1975 and 2007, the American Southwest suffered a water deficit in general (Seager et al. 2007). With this strong evidence of regional water deficits, it is surprising that there is no long-term trend of increasing salinity. It is possible that the trend is masked by oscillations on a smaller scale, or that extreme events are more common and droughts are simply followed by floods, and the long-term mean of the water balance is less important than the variance of the water balance.

Increasing sea surface temperatures along the Texas coast mimic what is occurring in many of the world's oceans. These increased surface temperatures are correlated to increased occurrence of intense tropical storms (Webster et al. 2005). Tropical storm activity varies from year to year, but more intense storms could have dramatic effects on coastal inundation and habitats.

Distribution of Mangroves

Mangroves are littoral plants that occur in tropical and subtropical coasts worldwide. These woody plants grow at the interface between land and sea, where they exist in conditions of high salinity, extreme tides, strong winds, high temperatures, and muddy, anaerobic soils (Kathiresan and Bingham 2001). There may be no other group of plants with such highly developed morphological and physiological adaptations to extreme conditions.

Mangrove habitats are among the world's richest repositories of biological diversity and primary productivity (Tomlinson 1986). Mangroves help maintain genetically diverse coastal fauna and flora that are of social-ecological value to humans throughout the world. They constitute an important source of benefits for humans by serving as coastal protection, habitat for other terrestrial and marine species, and nurseries for important commercial fisheries, including fish, mollusks, and crustaceans. In developing countries, mangrove forests constitute an important source of fuel, raw materials for construction, and food.

With more than 50 percent of the world's people living near the coast, mangroves constitute a critical element of the coastal environment. The main drivers of change in mangrove communities are competition for land for aquaculture, agriculture, infrastructure, and tourism. Some 15.2 million hectares of mangroves were estimated to exist worldwide in 2005, down from 18.8 million hectares in 1980 (FAO 2007). The drastic losses observed around the world have made mangrove conservation crucial not only for the biodiversity and human societies that depend on mangroves but also for the survival of coral reefs and seagrass beds (Spalding et al. 2009).

Six species of mangrove are reported to occur in the United States: *Rhizophora mangle*, *Laguncularia racemosa*, *Avicennia germinans*, *Conocarpus erectus*, *Avicennia schaueriana*, and *Acrostichum aureum* (FAO 2007). Most recent estimates indicate that there were 197,648 hectares of mangroves in 2001 in the United States. Together, the United States, Panama, and the Bahamas account for 82 percent of the total mangrove area in North and Central America; however, the second largest documented rate of loss in North and Central America was recorded in the United States. Over the past three decades, mangrove forests have experienced reductions of 1.3 percent from 1980 to 1990, 0.8 percent from 1990 to 2000, and 0.5 percent from 2000 to 2005 (FAO 2007).

Although mangroves are now under higher protection in areas such as the Everglades National Park in Florida, they are still being damaged mainly by drainage for agriculture, reclamation for urban development, and canalization in the United States. At a regional scale, hurricanes represent a serious threat to mangroves and cause significant losses in the United States (FAO 2007).

Black Mangrove

Four species of mangroves occur in the Gulf of Mexico: red mangrove (*Rhizophora mangle*), white mangrove (*Laguncularia racemosa*), black mangrove (*Avicennia germinans*), and button mangrove (*Conocarpus erectus*) (Sherrod and McMillan 1985). Among those, only black mangrove occurs in Texas, west of Apalachiocola Bay, Florida, and north to the Rio La Pesca, Tamaulipas, Mexico (Lot-Helgueras et al. 1975). Galveston Island is the historical northern limit of black mangroves (Sherrod and McMillan 1981), but the species appears to be establishing itself more firmly on the Louisiana coast (Twilley et al. 2001). Temperature, salinity, and several other environmental factors limit the distribution and survival of black mangrove in Texas. Black mangrove is the only species known to tolerate winters in Texas (Sherrod and McMillan 1981; Tunnell 2002). In Texas, mangroves are recognized as the only conspicuous native woody vegetation of the marsh–barrier island ecosystem (Judd 2002); they are sparsely distributed along tidal channels in bays and estuaries (Pulich and Scalan 1987; Withers 2002). This species rarely exceeds 6 feet in height, and it is most commonly distributed on the southern coast of Texas (Tunnell 2002).

The black mangrove has been impacted mainly by environmental and climatic fluctuations as well as anthropogenic modifications in Texas. Although human activity has resulted in the decrease of mangroves, it has also been responsible for the increase of intertidal marsh habitat within this past century because of coastal dredging and channelization (Pulich et al. 1997; Tremblay et al. 2008). Different climatic periods, however, appear to have had a large influence on mangrove population fluctuations during the past two centuries on the Texas coast (Sherrod and McMillan 1985).

On the Texas coast, there are four primary populations of black mangroves: Port Isabel (Cameron County), Harbor Island (Aransas Pass in Nueces and Aransas counties), Port O'Connor (Cavallo Pass in Calhoun County), and Galveston Island (Galveston County) (Sherrod and McMillan 1981).

Harbor Island contains one of the densest and largest populations of black mangroves documented since the 1930s on the Texas coast (Britton and Morton 1989). Harbor Island is a flood-tidal delta located at the mouth of Aransas Pass inlet, separating Mustang Island and San Jose Island (Fig. 4.2). Harbor Island also borders two bays, Aransas Bay to the north and Redfish Bay to the south. It is bounded by the Lydia Ann Channel, which connects the Intracoastal Waterway between Aransas Pass and Port Aransas. This tidal delta complex represents the most extensive and northernmost estuarine tropical wetland on the Texas coast (Pulich 2007).

Black mangroves are a good indicator for climate change because they occur in monospecific stands, are intertidal, and are at their northern limit for temperature on the Texas coast. The objective of the present study was to analyze the change in spatial cover over time of the Harbor Island population of black mangrove and its linkages with recent climate change patterns.

Changes in spatial coverage were analyzed by comparing past mapping efforts. A spatial information system was developed by digitizing available historical paper maps, visually interpreting digital aerial photographs, and integrating geographic information system (GIS) layers for the years 1930, 1979, 1995, and 2004. The data used and their sources were:

- 1930: Method: Visual interpretation of aerial photograph. Digitized from picture negative. Source: Sherrod 1980.
- 1979: Method: Visual interpretation of aerial photograph. Digitized from picture negative. Source: Sherrod 1980.
- 1995: Method: Visual interpretation of digital Ortho Quarterquads of color infrared photography. Horizontal resolution of 1 × 1 meter. Source: U.S. Geological Survey.
- 2004: Method: Image segmentation using blue, green, red, and near-infrared bands for each of the six processing areas. Classification of the habitat segments (as ESRI Shapefile format) was performed using CART analysis by the Coastal Services Center, National Oceanic and Atmospheric Administration (and consultant TerraSurv, Inc., of Pittsburgh, Pennsylvania), of color aerial photography (digital multispectral imagery ADS-40), scale 1:2,000; data ground truthed in May, June, and July 2006 and January 2007. Horizontal resolution of 2 × 2 meters. Source: National Agricultural Imagery Program.

Methods used in previous studies to determine mangrove abundance using aerial photography along the Texas coast were followed (Sherrod and McMillan 1981; Everitt and Judd 1989). Four individual maps, corresponding to each year analyzed, were compiled using polygons in Shapefile (vector model) in ArcGIS version 9.2 (ESRI) GIS software. Layers were georeferenced to the Universal Transverse Mercator (UTM zone 16) system and the

NAD83 Datum to allow spatial overlapping and comparisons. Maps of the mangrove expansion for each year were created using the same cartographic scale for comparison purposes.

The spatial coverage of black mangrove at Harbor Island during the four periods analyzed changed (Fig. 4.7). The area covered by black mangrove was calculated by summing the individual mangrove polygons (10.4 hectares each) on the island using GIS. There was a total increase in cover area of 131 percent over the 74-year period analyzed. The increase in area, however, was not linear during the period. The cover area decreased 47.16 percent between 1979 and 1995 but recovered by 2004, when it was greater than the previous extent in 1979.

The decrease of mangrove cover observed between 1979 and 1995 could be linked to the four years of freezes (i.e., the number of days with a minimum temperature below 32°F) that the Corpus Christi area experienced in 1982 (10 days), 1983 (13 days), 1985 (14 days), and 1989 (16 days) (Fig. 4.8). Mortalities of 85 percent of black mangrove were reported in 1984 and attributed to the freeze in 1983 (Sherrod and McMillan 1985). There were 11 years between 1940 and 1997 with more than 10 days below freezing in the Corpus Christi area. Of those, 81 percent of the freezes of 10 days or more together with the largest number of days (12–17 days) occurred since 1963.

The distribution pattern indicates that mangroves have occurred in most of the total area of Harbor Island, but especially in the southeastern portion. The total cover on the island, extrapolated from the study area, is 2,484.43 hectares in 2004.

Although a small increase in air temperature may have little direct effect on mangrove physiology, it is expected to cause a northern shift in the freeze line and consequently a northern shift in the distribution of mangroves (Field

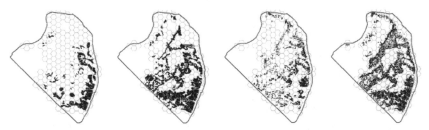

FIGURE 4.7.   Area of Harbor Island, Texas, covered by black mangrove in each of four years. *From left to right:* 1930: 24.9% (235.54 ha); 1979: 58% (548.29 ha); 1995: 39.4% (372.56 ha); 2004: 57.7% (545.58 ha). Percentages equal maximum combined historical coverage up to 2004. Hexagons represent maximum combined historical extent up to 2004. Outer boundary represents Harbor Island limits in 2004.

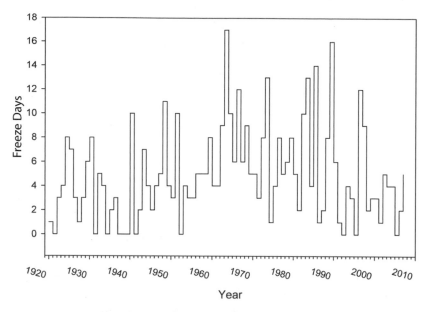

FIGURE 4.8. Number of days with minimum temperature below 0°C (32°F), 1920–2007. Data source: Corpus Christi International Airport, Texas; Station COOP ID 412015, NCDC/NOAA.

1995; Twilley et al. 2001). Black mangroves appear to be more firmly established on the Louisiana coast (Twilley et al. 2001). Projections presented here show an increase in mangrove distribution along the south-central Texas coast.

Mangroves represent a well-defined niche in coastal zonation and therefore are likely to be early indicators of the effects of global climate change. Warmer temperatures, as predicted by the United Nations Intergovernmental Panel on Climate Change (IPCC 2007), will facilitate the expansion of tropical species northward, such as mangroves (Montagna et al. 2007). Black mangroves, however, are typically found in sheltered coasts of the intertidal fringe and as such are sensitive to temperature and inundation. Thus, climate change has other implications for mangrove populations, including an increase in sea level, a change in the number of days below freezing temperatures, and changes in the frequency and intensity of hurricanes. These changes are expected to have a greater impact on the Texas coast (Field 1995; Sherrod and McMillan 1985; Montagna et al. 2007; Ning et al. 2003; Tremblay et al. 2008).

In Texas, black mangrove is at the northern limit of its range, where catastrophic mortality due to freezes may occur. Although McMillan (1975) suggested that the Texas population belongs to a genetic race capable of surviving colder temperatures, the December 1983 freeze resulted in an 85 percent reduction in the Texas population (Sherrod and McMillan 1985). The results of the Harbor Island analysis indicate that recovery could take more than 20 years.

Mangrove communities occur between high and low tides, and elevation of the shoreline due to sea-level rise would leave mangroves in deeper waters and exposed to erosion from wave action. In addition, sea-level rise will increase water volume, which could alter the salinity regime of bays and estuaries in Texas (Montagna et al. 2007). Ning et al. (2003) have reported that a 20-inch rise in sea level would cause large losses of mangroves in southwest Florida (where most of the mangroves of the United States occur).

Although climate change is expected to cause an increase of 3–7°C (5.4–12.6°F) in the Gulf of Mexico region (Twilley et al. 2001), it is important to note that mangrove populations are affected by anthropogenic disturbances as well. Most of the changes at Harbor Island could have been caused by dredging and channel construction. Intense dredging activities occurred between 1958 and 1975 (Pulich et al. 1997). Particularly, the southwest corner of the island has been subject to dredging and channelization due to oil and gas exploration and to disposal of dredge material excavated from the Corpus Christi Ship Channel. It has also been reported that nearshore areas that were covered by vegetation in 1975 were buried by a discharge of dredge material in 1994.

Because of their location in the front line between ocean and land, mangroves in the Gulf of Mexico also face natural threats such as hurricanes. The area of mangrove forest impacted by hurricanes in Texas has not been quantified. In Mexico, however, 10,000–14,000 hectares have been lost annually because of hurricane impacts since 1980 (FAO 2007).

Red Mangroves

Isolated red mangroves have been observed in a few places on the southern coast of Texas since 1983 (Tunnell 2002). Recent hurricane activity appears to be facilitating the invasion of the red mangrove (*Rhizophora mangle*) on the Texas coast, judging from the observation of seeds on Gulf beaches. The red mangrove individuals established in bays between South Padre Island and Matagorda Island have increased since the intense hurricane season of 2005 (Montagna et al. 2007). The proximity of Harbor Island to Aransas Pass, the Gulf inlet, explains the recent appearance of red mangrove there. Recently, Zomlefer et al. (2006) documented the most northern occurrence of red mangrove, at latitude 29 degrees 42.94' N in St. Johns County, on the Atlantic coast of Florida. These findings suggest that all mangrove species are expanding their range northward.

CONCLUSIONS

The Texas coast may be one of the best places on Earth to observe climate change effects, outside the polar regions, because the major physical drivers, such as temperature, rainfall, and sea-level rise, can all have large and interacting effects in the northwestern Gulf of Mexico, and indicator species for climate change effects exist to illustrate the story. Two direct effects are already observable in the instrumental record: rapid sea-level rise and rising sea temperatures. The rates of sea-level rise are high because of subsidence, which causes the relative rise to be that much greater. The increasing temperatures are already manifesting indirect changes in habitats and water quality.

Black mangroves, which are sensitive to freezes, are expanding northward. Species that are even more sensitive to cold, such as the red mangrove, are showing up on the Texas coast. Rapid sea-level rise, however, may interact with habitat change to alter the trajectory of succession of coastal landscapes. It is not clear exactly what will happen. One possibility is that the rising sea level will simply drown wetland habitats, but as long as plant growth and soil stabilization by plant roots occur at a rate higher than apparent sea-level rise, the habitats can simply move with moving shorelines. There is little reason to conclude that shorelines will not change.

Water quality change may be the most pernicious change of all, even though it is an indirect change driven by the lower solubility of oxygen in warmer water. The potential for hypoxia (low concentrations of dissolved oxygen) is very great and increasing. Called dead zones by the news media, hypoxic areas are known to be large and expanding in number, extent, and duration. Hypoxia is very destructive to coastal ecosystems, leading to lower biomass, productivity, and diversity, and it can alter food webs such that desirable species can no longer be produced in an area. Although hypoxia is caused by excess loading of nutrients from watersheds to coastal waters, it is clear that physical processes also play a role in lowering dissolved oxygen concentrations.

Earlier studies focused on how rainfall and consequent changes in freshwater inflow might alter systems, but there is no evidence in the recent instrumental record that salinities are changing along the Texas coast. Focus should be placed on adaptation to hydrological changes in climate. This would include better coastal planning so that human activities can account for changing coastlines and habitats, and more concern about nutrient reductions. If climate change drives down dissolved oxygen concentrations, then the only recourse to adapt to this condition will be to put further controls on nutrient additions to coastal waters.

ACKNOWLEDGMENTS
The authors thank Nicole Selly, who digitized the mangrove distribution maps, and William L. Longley, who wrote the introductory sections "Texas Estuaries" and "River Basins" as part of the 1995 edition of this book.

REFERENCES

Anderson, J. B., 2007. *Formation and Future of the Upper Texas Coast.* Texas A&M University Press, College Station.

Applebaum, S., P. A. Montagna, and C. Ritter, 2005. Status and Trends of Dissolved Oxygen in Corpus Christi Bay, Texas, USA. *Environmental Monitoring and Assessment* 107:297–311.

Britton, J. C., and B. Morton, 1989. *Shore Ecology of the Gulf of Mexico.* University of Texas Press, Austin.

Brown, L. F. Jr., J. H. McGowen, T. J. Evans, C. G. Groat, and W. L. Fisher, 1977. Environmental Geologic Atlas of the Texas Coastal Zone: Kingsville Area. Bureau of Economic Geology, University of Texas, Austin.

Brown, L. F. Jr., J. L. Brewton, T. J. Evans, J. H. McGowen, W. A. White, C. G. Groat, and W. L. Fisher, 1980. Environmental Geologic Atlas of the Texas Coastal Zone: Brownsville-Harlingen Area. Bureau of Economic Geology, University of Texas, Austin.

Bruun, P., 1962. Sea-level Rise as a Cause of Shore Erosion. *Journal of the Waterways and Harbors Division* 88:117–130.

Davis, R. A. Jr., W. G. Fingleton, G. R. Allen Jr., C. D. Crealese, W. M. Johanns, J. A. O'Sullivan, C. L. Reive, S. J. Scheetz, and G. L. Stranaly, 1973. Corpus Christi Pass: A Hurricane Modified Tidal Inlet on Mustang Island, Texas. *Contributions in Marine Science* 17:123–131.

Diaz, R. J., and R. Rosenberg, 2008. Spreading Dead Zones and Consequences for Marine Ecosystems. *Science* 321:926–929.

Douglas, B. C., 1991. Global Sea Level Rise. *Journal of Geophysical Research* 96:6981–6992.

Emanuel, K., 2005. Increasing Destructiveness of Tropical Cyclones over the Past 30 Years. *Nature* 436:686–688.

Everitt, J. H., and F. W. Judd, 1989. Using Remote-sensing Techniques to Distinguish and Monitor Black Mangrove (*Avicennia germinans*). *Journal of Coastal Research* 5:737–745.

FAO, 2007. The World's Mangroves, 1980–2005. Forestry Paper 153. Food and Agriculture Organization of the United Nations, Rome.

Field, C. D., 1995. Impact of Expected Climate Change on Mangroves. *Hydrobiologia* 295:75–81.

Fisher, W. L., J. H. McGowen, L. F. Brown Jr., and C. G. Groat, 1972. Environmental Geologic Atlas of the Texas Coastal Zone: Galveston-Houston Area. Bureau of Economic Geology, University of Texas, Austin.

Gibeaut, J. C., and T. A. Tremblay, 2003. *Coastal Hazards Atlas of Texas: A Tool for Hurricane Preparedness and Coastal Management,* Vol. 3. Bureau of Economic Geology, University of Texas, Austin.

Gibeaut, J. C., W. A. White, R. C. Smyth, J. R. Andrews, T. A. Tremblay, R. Gutiérrez, T. L. Hepner, and A. Neuenschwander, 2003. Topographic Variation of Barrier Island Subenvironments and Associated Habitats: Coastal Sediments '03, Crossing Disciplinary Boundaries. Volume CD-ROM. Proceedings, Fifth International Symposium on Coastal Engineering and Science of Coastal Sediment Processes, Clearwater Beach, Fla.

IPCC, 2001. *Climate Change 2001: The Scientific Basis.* Contribution of Working Group 1 to the Third Assessment Report of the Intergovernmental Panel on Climate Change. Cambridge University Press, Cambridge, U.K., and New York.

IPCC, 2007. *Climate Change 2007: The Physical Science Basis.* Contribution of Working Group 1 to the Fourth Assessment Report of the Intergovernmental Panel on Climate Change. Cambridge University Press, Cambridge, U.K., and New York.

Judd, F. W., 2002. Tamaulipan Biotic Province. Pp. 38–58 in: *The Laguna Madre of Texas and Tamaulipas.* J. W. Tunnell Jr. and F. W. Judd (eds.). Texas A&M University Press, College Station.

Kathiresan, K., and B. L. Bingham, 2001. Biology of Mangroves and Mangrove Ecosystems. *Advances in Marine Biology* 40:81–251.

Larkin, T. J., and G. W. Bomar, 1983. *Climatic Atlas of Texas.* Texas Department of Water Resources, Austin.

Longley, W. L., 1994. Freshwater Inflows to Texas Bays and Estuaries: Ecological Relationships and Methods for Determination of Needs. Joint Estuarine Research Study, Texas Water Development Board and Texas Parks and Wildlife Department, Austin.

Longley, W. L., 1995. Estuaries. Pp. 88–118 in: *The Impact of Global Warming on Texas: A Report of the Task Force on Climate Change in Texas.* G. R. North, J. Schmandt, and J. Clarkson (eds.). University of Texas Press, Austin.

Lot-Helgueras, A., C. Vazquez-Yanez, and F. Menendez, 1975. Physiognomic and Floristic Changes near the Northern Limit of Mangroves in the Gulf Coast of Mexico. Pp. 52–61 in: *Proceedings of the International Symposium on the Biology and Management of Mangroves,* Vol. 1. E. Walsh, S. C. Snedaker, and H. J. Teas (eds.). University of Florida, Gainesville.

McGowen, J. H., C. V. Procter Jr., L. F. Brown Jr., T. J. Evans, W. L. Fisher, and C. G. Groat, 1976. Environmental Geologic Atlas of the Texas Coastal Zone: Port Lavaca Area. Bureau of Economic Geology, University of Texas, Austin.

McMillan, C., 1975. Adaptive Differentiation to Chilling in Mangrove Populations. Pp. 62–68 in: *Proceedings of the International Symposium on the Biology and*

Management of Mangroves, Vol. 1. E. Walsh, S. C. Snedaker, and H. J. Teas (eds.). University of Florida, Gainesville.

Miller, L., and Douglas, B. C., 2004. Mass and Volume Contributions to Twentieth-century Global Sea Level Rise. *Nature* 428:406–409.

Montagna, P. A., J. C. Gibeaut, and J. W. Tunnell Jr., 2007. South Texas Climate 2100: Coastal Impacts. Pp. 57–77 in: *The Changing Climate of South Texas, 1900–2100: Problems and Prospects, Impacts and Implications*. J. Norwine and K. John (eds.). Texas A&M University, Kingsville.

Morton, R. A., and J. G. Paine, 1984. *Historical Shoreline Changes in Corpus Christi, Oso, and Nueces Bays, Texas Gulf Coast*. Bureau of Economic Geology, University of Texas, Austin.

Morton, R., J. Bernier, and J. Barras, 2006. Evidence of Regional Subsidence and Associated Interior Wetland Loss Induced by Hydrocarbon Production, Gulf Coast Region, USA. *Environmental Geology* 50:261–274.

Ning, Z. H., R. E. Turner, T. Doyle, and K. Abdollahi, 2003. *Preparing for a Changing Climate: The Potential Consequences of Climate Variability and Change, Gulf Coast Region*. U.S. Environmental Protection Agency, GCRCC, and Louisiana State University, Baton Rouge.

Orlando, S. P. Jr., L. P. Rozas, G. H. Ward, and C. J. Klein, 1993. Salinity Characteristics of Gulf of Mexico Estuaries. National Oceanic and Atmospheric Administration, Office of Ocean Resources Conservation and Assessment, Silver Spring, Md.

Overpeck, J. T., B. L. Otto-Bliesner, G. H. Miller, D. R. Muhs, R. B. Alley, and J. T. Kiehl, 2006. Paleoclimatic Evidence for Future Ice-sheet Instability and Rapid Sea-level Rise. *Science* 311:1747–1750.

Paine, J. G., 1993. Subsidence of the Texas Coast: Inferences from Historical and Late Pleistocene Sea Levels. *Tectonophysics* 222:445–458.

Paine, J. G., and R. A. Morton, 1993. *Historical Shoreline Changes in Copano, Aransas, and Redfish Bays, Texas Gulf Coast*. Bureau of Economic Geology, University of Texas, Austin.

Pritchard, D. W., 1967. What Is an Estuary: Physical Viewpoint. Pp 3–5 in: *Estuaries*. Publ. 83, American Association for the Advancement of Science, Washington, D.C.

Prouty, J. S., and D. B. Prouty, 1989. Historical Back Barrier Shoreline Changes, Padre Island National Seashore, Texas. *Gulf Coast Association of Geological Societies, Transactions* 39:481–490.

Pulich, W. Jr., 2007. Texas Coastal Bend. In: *Seagrass Status and Trends in the Northern Gulf of Mexico, 1940–2002*. L. Handley, D. Altsman, and R. DeMay (eds.). U.S. Geological Survey Scientific Investigations Report 2006–5287, U.S. Environmental Protection Agency 855-R-04-003, Reston, Va.

Pulich, W. Jr., and R. S. Scalan, 1987. Organic Carbon and Nitrogen Flow Marine Cyanobacteria to Semiaquatic Insect Food Webs. *Contributions in Marine Science* 30:27–37.

Pulich, W. Jr., C. Blair, and W. A. White, 1997. Current Status and Historical Trends of Seagrass in the Corpus Christi Bay National Estuary Program Study Area.

Publication CCBNEP-20. Natural Resource Conservation Commission, Austin, Tex.

Ratzlaff, K. W., 1980. *Land-surface Subsidence in the Texas Coastal Region.* U.S. Department of Interior Geological Survey, Austin, Tex.

Rignot, E., and P. Kanagaratnam, 2006. Changes in the Velocity Structure of the Greenland Ice Sheet. *Science* 311:986–990.

SAS Institute, 2004. *SAS/STAT 9.1 User's Guide.* SAS Institute, Inc., Cary, N.C.

SAS Institute, 2006. *SAS 9.1.3 Language Reference: Dictionary,* Vols. 1–4. 5th ed. SAS Institute, Inc., Cary, N.C.

Seager, R., M. Ting, I. Held, Y. Kushnir, J. Lu, G. Vecchi, H. P. Huang, N. Harnik, A. Leetmaa, N. C. Lau, C. Li, J. Velez, and N. Naik, 2007. Model Projections of an Imminent Transition to a More Arid Climate in Southwestern North America. *Science* 316:1181–1184.

Sharp, J. M. Jr., S. J. Germiat, and J. G. Paine, 1991. Re-evaluation of the Causes of Subsidence along the Texas Gulf of Mexico Coast and Some Extrapolation of Future Trends. IAHS, *Fourth International Symposium on Land Subsidence* 200:397–405.

Shepherd, A., and D. Wingham, 2007. Recent Sea-level Contributions of the Antarctic and Greenland Ice Sheets. *Science* 315:1529–1532.

Sherrod, C. L., 1980. Present and Past Distribution of Black Mangrove (*Avicennia germinans* [L.] L.) on the Texas Gulf Coast. Thesis (M.A.), University of Texas, Austin.

Sherrod, C. L., and C. McMillan, 1981. Black Mangrove, *Avicennia germinans,* in Texas: Past and Present Distribution. *Contributions in Marine Science* 24:115–131.

Sherrod, C. L., and C. McMillan, 1985. The Distributional History and Ecology of Mangrove Vegetation along the Northern Gulf of Mexico Coastal Region. *Contributions in Marine Science* 28:129–140.

Spalding, M., M. Kainuma, and L. Collins, 2009. *World Atlas of Mangroves.* Earthscan, U.K.

Texas Water Data Services, 2008. Water Quality Data. Coastal Fisheries Division, Texas Parks and Wildlife Department. http://data.crwr.utexas.edu/

Tolan, J. M., 2007. El Niño-Southern Oscillation Impacts Translated to the Watershed Scale: Estuarine Salinity Patterns along the Texas Gulf Coast, 1982 to 2004. *Estuarine Coastal and Shelf Science* 72:247–260.

Tomlinson, P. B., 1986. *The Botany of Mangroves.* Cambridge Tropical Biology Series. Cambridge University Press, U.K.

Tremblay, T. A., J. S. Vincent, and T. R. Calnan, 2008. Status and Trends of Inland Wetland and Aquatic Habitat in the Corpus Christi Area. Coastal Bend Bays and Estuaries Program, Publication 55, Corpus Christi.

Tunnell, J. W. Jr., 2002. The Environment. Pp. 73–84 in: *The Laguna Madre of Texas and Tamaulipas.* J. W. Tunnell Jr. and F. W. Judd (eds.). Texas A&M University Press, College Station.

Twilley, R. R., E. J. Barron, H. L. Gholz, M. A. Harwell, R. L. Miller, D. J. Reed, J. B. Rose, E. H. Siemann, R. G. Wetzel, and R. J. Zimmerman, 2001. *Confronting*

Climate Change in the Gulf Coast Region. Union of Concerned Scientists and Ecological Society of America, Cambridge, Mass.

Webster, P. J., G. J. Holland, J. A. Curry, and H. R. Chang, 2005. Changes in Tropical Cyclone Number, Duration, and Intensity in a Warming Environment. *Science* 309:1844–1846.

Wentz, F. J., L. Ricciardulli, K. Hilburn, and C. Mears, 2007. How Much More Rain Will Global Warming Bring? *Science* 317:233–235.

White, W. A., R. A. Morton, R. S. Kerr, W. D. Kuenzi, and W. B. Brogden, 1978. *Land and Water Resources, Historical Changes, and Dune Criticality: Mustang and North Padre Islands, Texas.* Bureau of Economic Geology, University of Texas, Austin.

White, W. A., R. A. Morton, and C. W. Holmes, 2002. A Comparison of Factors Controlling Sedimentation Rates and Wetland Loss in Fluvial-deltaic Systems, Texas Gulf Coast. *Geomorphology* 44:47–66.

White, W. A., T. A. Tremblay, R. L. Waldinger, and T. R. Calnan, 2006. *Status and Trends of Wetland and Aquatic Habitats on Texas Barrier Islands Coastal Bend.* Final Report NA04NOS4190058. Coastal Coordination Division, Texas General Land Office, Austin.

Williams, H. F. L., 1999. Sand-spit Erosion Following Interruption of Longshore Sediment Transport: Shamrock Island, Texas. *Environmental Geology* 37:153–161.

Withers, K., 2002. Wind-tidal Flats. Pp. 114–126 in: *The Laguna Madre of Texas and Tamaulipas.* J. W. Tunnell Jr. and F. W. Judd (eds.). Texas A&M University Press, College Station.

Zervas, C., 2001. Sea Level Variations of the United States, 1854–1999. National Oceanic and Atmospheric Administration, Technical Report NOS CO-OPS 36. Silver Spring, Md.

Zimmerman, R. J., T. J. Minello, E. F. Klima, and J. M. Nance. 1991. Effects of Accelerated Sea-level Rise on Coastal Secondary Production. Pp. 110–124 in: *Coastal Wetlands Zone '91 Conference.* H. S. Bolton and O. T. Magoon (eds). ASCE, Long Beach, Calif.

Zomlefer, W. B., W. S. Judd, and D. E. Giannasi, 2006. Northernmost Limit of *Rhizophora mangle* (red mangrove; Rhizophoraceae) in St. Johns County, Florida. *Castanea* 71:239–244.

CHAPTER 5

Biodiversity

Jane M. Packard, Wendy Gordon, and Judith Clarkson

The concept of biodiversity refers to (1) the number of types of natural ecological regions (ecoregions), (2) the number of native species within each ecoregion, and (3) the genetic variation within species (Janetos et al. 2008; Siikamaki 2008). For example, within the United States, vertebrate species richness naturally tends to be highest in southern areas and decreases with seasonally harsh winters in the north. Large states with boundaries that encompass a diverse array of ecosystems tend to contain a greater number of species than small homogeneous states. For this reason, Texas is second only to California in number of total species (Siikamaki 2008; Texas Environmental Profile 2008). Genetic aspects of biodiversity are illustrated by the global hotspot of endemism found in the isolated springs and cave systems of the Edwards Plateau, a natural legacy unique to Texas.

Twelve major natural regions (ecoregions) are currently recognized within Texas (Fig. 5.1), ranging from deserts to prairies and pine forests, from mountains to coastal marshes (Griffith et al. 2007). Within each ecoregion, the landscape is further subdivided into vegetation types that correspond to variations in land and water features. This biologically diverse landscape represents a natural living library for scientists tracking the changes associated with climate change. Across the continent, the effects of climate change include not only redistribution of resident species but also the timing (phenology) of reproduction in resident species and arrival of migrant species (Janetos et al. 2008). This reshuffling of biotic communities is likely to result in ripples up and down the food chain, which are difficult to predict with the limited information currently available.

Projecting impacts of climate change on biodiversity is a challenge for scientists and decision makers (Powledge 2008). Systematic scientific analysis of the problem is lacking for the state of Texas, so in this chapter we will review hypotheses derived from national and global studies (Janetos et al. 2008) while emphasizing the need for a more in-depth approach to adaptive management of biodiversity in Texas based on strategic monitoring and research. To be consistent with other chapters, this preliminary application of theory will be based on the range of scenarios of climate change projected for Texas (Chapter 1). Trends projected from historical data, however,

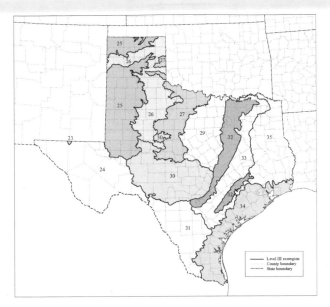

FIGURE 5.1. The 12 major natural regions of Texas. The historically rich biological diversity in Texas can be classified into 12 ecoregions, identified here by their names and numbers within the Environmental Protection Agency's national system: Arizona–New Mexico Mountains (23), Chihuahuan Desert (24), High Plains (25), Southwestern Tablelands (26), Central Great Plains (27), Cross Timbers (29), Edwards Plateau (30), Southern Texas Plains (31), Texas Blackland Prairies (32), East Central Texas Plains (33), Western Gulf Coastal Plains (34), South Central Plain (35). These regions have been further subdivided by ecologists according to distinctive vegetation and geology within each region (Griffith et al. 2007). Some educational sources report only 11 ecoregions (Texas Environmental Profile 2008), deleting the Arizona–New Mexico mountain habitat because it is represented by only one peak in the Guadalupe Mountains.

have not always been consistent with projections of global climate models (Chapter 2). This discussion will show that many other factors, such as aquatic systems (Chapters 3 and 4) and growth of human populations (Chapters 6 and 7), probably will interact with climate in determining changes in the distribution of native flora and fauna across Texas. In particular, uncertainties associated with changes in human populations, land use, water use, seasonal rainfall, the rate of climate change, and climatic variability (e.g., storms, droughts, and freezes) make it very difficult to make statewide projections about changes in biodiversity.

Therefore, this chapter first presents background information about known threats to biodiversity and uncertainties regarding projected changes. Second, we outline some general principles about the potential effects of climate change on biodiversity. Third, we describe the natural regions of Texas to illustrate both the richness of the natural heritage of the state and the reasons ecological changes would be expected to differ for each region. Predicting exactly how these changes are likely to occur would require a regional analysis beyond the scope of this chapter. Finally, at the conclusion of the chapter, we recommend the types of studies and strategies needed to manage biodiversity in Texas. The urgent need for decision makers to support an integrated and systematic research program will be made very evident.

BIODIVERSITY AND TEXAS

According to the NatureServe database (Stein 2002), Texas supports the highest number of reptiles (149 species) and birds (635 species) of all the states in the nation and the second highest number of plants (6,313 species) and mammals (184 species). Still in the process of counting, the inventory of insect species may be as high as 29,000 species (TPWD 2005). With the third highest rate of endemism, Texas is ranked second only to California in overall estimates of biodiversity (Stein 2002). Texas ranks fourth in the United States in terms of numbers of known species lost to extinction.

This rich natural heritage of Texas is under threat from a variety of factors likely to interact with climate change (Table 5.1). According to records of known species extinctions, approximately 0.2–0.4 percent of all described U.S. species have gone extinct; for birds this number is closer to 5 percent (Siikamaki 2008). Ecological theory suggests that several factors contribute to the high extinction probability of certain types of species (Siikamaki 2008). In general, characteristics of species that are most vulnerable to extinction include (a) large body size, (b) small population size or range, (c) adaptations that have evolved in isolation or with an evolutionary history of infrequent disturbances, (d) limited dispersal or colonization mechanisms, (e) migratory cycles, (f) colonial breeding or nesting, and (g) dependence on vulnerable species lower in the food chain.

Threats to Biodiversity in Texas

The adaptive resilience of some species native to Texas will likely be overwhelmed by the cumulative effects of multiple stressors, especially when combined with additional barriers to movement as human populations expand (Packard 1995; Packard and Cook 1995). For example, Houston toads are an endangered endemic species currently on the brink of extinction in Texas despite policies favoring their protection. Native habitat for this species has been displaced by urban and agricultural expansion. Increased

TABLE 5.1. Factors likely to interact with climate change to threaten biodiversity in Texas[a]

CAUSE/STRESSOR	DRIVERS CONTRIBUTING TO RATE OF CHANGE IN STRESSORS	EXAMPLES OF PREDICTED CUMULATIVE EFFECTS OF STRESSORS ON ECOSYSTEMS
Global climate change	• Release of carbon through deforestation • Atmospheric increase of carbon due to burning of fossil fuels • Release of chemicals affecting the ozone layer	• The accelerated rate of change will be faster than abilities of native species to adapt. • Optimal conditions for species will shift to locations outside protected areas. • Productive agricultural land use will move. • Sea level will rise, flooding marshes. • The frost-line will move toward the poles and to higher elevations; winters will be shorter. • Rainfall will be more unpredictable with changes in tropical storms and drought cycles (hurricanes and El Niño oscillations).
Habitat destruction and degradation	• Vegetation removal • Urban and agricultural development • Suburban sprawl	• Carbon sequestration will decrease. • Habitat will become more fragmented, in smaller disconnected acreages. • Connectivity between habitat fragments will be blocked by roads and urbanization. • Carrying capacity of habitats will change, decreasing for specialized species and increasing for colonizing species.
Water pollution and harvest	• Urban, agricultural, and industrial sources release bacteria, fertilizers, chemical waste, and heavy metals into aquatic systems • Demand for water affects aquifer depletion, water control, capture, storage, and diversion projects	• Algal blooms will negatively impact nursery habitat for fish and other aquatic species. • Flood cycles in rivers will change, affecting channel scouring and sediment deposits. • Reservoirs from new dams will flood riparian habitat (e.g., bottomland hardwoods). • Aquifer-fed springs will dry up. • Water flow in rivers flowing into estuaries will be influenced by transfer between basins, discharge from urban areas, and changes in soil permeability. • Saltwater intrusion will increase. • Wetland soil moisture cycles will change.
Invasive species	• Increased transport in a global market • Disturbed soils and plants	• Food chains will change as species are released from (or encounter) new predators, diseases, parasites, and disturbances. • Colonizing species will outcompete native species typical of old-growth succession.

[a] In addition to the direct effects of global climate change, changes in biodiversity will occur as a result of indirect interactions with other components of ecosystems, due to stressors driven both by human activities and by other organisms (Brown et al. 2001; IPCC 2007; Griffith et al. 2007).

risk of drought during the reproductive seasons of Houston toads is likely to reduce the capacity for the population to bounce back when hit by predators or disease (Swannack et al. 2009). Through private-public partnerships, genetic stocks are currently maintained in captivity, anticipating future restoration strategies.

Aquatic systems also face significant threats in Texas because of increasing water demand from a burgeoning human population, and climate change is likely to exacerbate the threats (Texas Water Development Board 2007). New ground and surface water supply projects are on tap, which through direct and indirect effects will likely diminish the availability and quality of existing aquatic habitat. For example, higher temperatures will directly increase water temperatures in lakes, wetlands, and rivers, resulting in decreased dissolved oxygen concentrations. Rates of decay will accelerate, possibly leading indirectly to eutrophication and more frequent algal blooms. In addition to increasing evaporative losses, higher temperatures could prove lethal to some species or significantly disrupt their growth and reproduction. Changes in the seasonality of river flows and in the amount and distribution of rainfall could alter magnitude, duration, frequency, timing, and rate of change in river flow, which could adversely affect riverine, estuarine, and riparian species adapted to specific flow regimes for spawning cues, regeneration, or other physiological processes.

General Trends on a Scale Larger than Texas

The interactions of multiple factors make it very difficult to assign responsibility for loss of biodiversity directly to climate change (IPCC 2007). Species that are already under stress because of other factors will be more sensitive to the effects of higher temperatures and possibly reduced precipitation. The following trends, however, are already becoming evident across the North American continent (IPCC 2007; Janetos et al. 2008):

- Rising temperatures are resulting in the lengthening of growing seasons and changing migration patterns of birds and butterflies.
- Coastal and nearshore ecosystems are vulnerable to increases in air and water temperature, as well as changes in freshwater inflows, saltwater intrusion from rising sea level, nutrient enrichment, oxygen depletion, acidification, and oscillating current patterns such as El Niño.
- Pests and diseases are increasing in range because warmer winters reduce die-off, and parasite development rates and infectivity increase with temperature.
- Woody shrubs invading prairie grasslands are favored by increases in concentrations of carbon dioxide, changes in soil moisture cycles, fire suppression, and soil disturbances.

- Disturbances created from interaction of drought, pests, diseases, and fire can result in dramatic changes in ecosystems. Trees that are stressed by a multiyear drought may be too stressed to fight off insect infestations, such as southern pine bark beetle.

Because species are predicted to be differentially affected by climate change and their rates of adaptation may differ significantly from one another, it is expected that some ecosystems or species assemblages may be disrupted. For instance, MacMynowski and Root (2007) have found, in a study of 127 migratory bird species over 20 years in the migratory flyway through the central United States, that short-range migrants typically respond to temperature alone, which seems to correlate with food supply, but long-range migrants respond more to variation in the overall climate system. There is no reason to expect migrants and their respective food sources to shift their distribution at the same rate, potentially leading to mistimed reproduction in species at risk, such as the golden-cheeked warbler in Texas.

Uncertainty Regarding Projections

Although decision makers want clear-cut projections about the potential effects of climate change on biodiversity, most scientists are very cautious about the statements they make because ecological systems are more sensitive to the frequency and duration of climatic extremes than to the mean temperature. Furthermore, many projected ecological changes are expected to be nonlinear (Burkett et al. 2005). So far, global climate models do not provide the precision needed to project the local duration, frequency, or seasonality of precipitation. Therefore, ecologists currently do not have the information they need to make accurate projections. This information gap stands in the way of science-based decision making on conservation issues.

Furthermore, it is difficult to distinguish the potential influence of climate change from other interacting factors, such as land-use change. For example, the shift from grasses to shrubs in the Chihuahuan Desert may also be explained by overgrazing, which reduces soil coverage (Schlesinger et al. 1990) and increases shrub seed dispersal (Archer 1990). In addition, as noted above, species are likely to exhibit unique responses to climate change. The relative competitive abilities of plant species that differ in photosynthetic pathways (i.e., C_3 vs. C_4) may change due to interactions of factors such as carbon dioxide stimulation, warmer winter temperatures, and increased summer precipitation. The interrelated nature of the hydrologic cycle and terrestrial vegetation has been the subject of much experimentation in an effort to tease out winners and losers among plant species in a carbon dioxide–enriched world (Gordon and Huxman 2007). Unfortunately, there is still much uncertainty about these interactions, especially in water-limited systems.

Interaction of Species Traits and Climatic Shifts

The differential response of species to climate change is likely to result in a "reshuffling of the deck" (Fox 2007). For example, paleoecologists have documented no-analog communities, meaning combinations of species that existed under prior climatic regimes but no longer exist in the same combination today. This historical pattern of plant communities is evidence of the disparate effects of climate shifts on individual species and suggests that ecological communities of the future may be dissimilar from those of today (Williams and Jackson 2007), with current biodiversity hotspots most at risk (Callicott et al. 2007; Williams et al. 2007).

Imagine a cross superimposed on the map of Texas. The horizontal bar is the freeze line, and the vertical bar is the tree line (threshold of soil moisture retention for trees). Now, move that cross to the northeast and you have a simplistic image of how species distributions are projected to change in Texas with an increase in temperature, assuming (1) no change in precipitation, (2) biotic communities move as a unit, (3) no complications due to soil requirements, and (4) no potential barriers such as rivers, escarpments, highways, or urban areas. Reality differs from this simplistic model according to local variation in the interactions between geographical features and human activities. Geographical features include bands of soil, rivers, canyons, mountains, and escarpments that may influence connectivity (e.g., species' movements between habitat fragments). Depending on dispersal mechanisms, individual species are predicted to respond differently to potential geographic barriers. Overlaid on the geographic diversity are the added effects of growing human populations.

It quickly becomes apparent that in-depth analyses of the landscapes in each region are needed to clarify the projected effects of climate change that threaten biodiversity in Texas. This is particularly true if projected changes in agricultural practices, suburban expansion, and urban development are to be taken into account (Callicott et al. 2007; Acevedo et al. 2008). Precedents exist for these types of analyses in other regions. For example, population biologists and modelers of species distribution were brought together by a collaborative program in Europe to explore a new framework for integrating the influence of climate and land-use change on population dynamics (Barnard and Thuiller 2008). In response to an executive order in 2005,

California initiated the Scenarios Project to investigate projected impacts of climate change on six sectors including ecosystems (Cayan et al. 2008). In the Pacific Northwest, neighboring states have participated in a scientific synthesis to support adaptation to effects of climate change in the Puget Sound ecosystem (Ruckelshaus and McClure 2007).

GENERAL PRINCIPLES: ECOLOGY AND CLIMATE CHANGE

In this section, we elaborate on the intellectual framework used to address threats to biodiversity in regions outside Texas, with the goal of applying these ideas to specific Texas ecoregions in the subsequent section. Four general principles are recognized regarding the effects of climate change on ecological systems (IPCC 2007):

1. Climate is the key determining variable of species distributions. As the earth warms, species tend to shift to northern latitudes and higher altitudes.

2. Loss of biological diversity is likely to result from the interaction of climate change and other human-induced stressors (Table 5.1).

3. The distribution of species vulnerable to extinction depends on their genetic plasticity (or ability to adapt to new environmental conditions in situ), their intrinsic ability to migrate, the existence of continuous habitat along gradients of climate change, and both the amount and rate of climate change.

4. Ecosystems are inherently complex, requiring sophisticated models to project changes in biodiversity related to alternative scenarios of climate change, including interactive effects of multiple factors (e.g., temperature, precipitation, evaporative demand, soil and substrate, soil moisture retention, physiological adaptations, and relations between species).

These four general principles are explained below to establish a conceptual framework. They will then be illustrated with examples.

Shift to Higher Latitudes and Altitudes

The movement of species in regions of North America in response to climate warming is expected to result in shifts of species ranges poleward and upward along elevational gradients in mountain regions (Parmesan 2006). Over the past two decades and across temperate latitudes, scientists have measured increases in summer photosynthetic activity and the amplitude of the annual carbon dioxide cycle, as well as earlier spring green-up, by as much as 10–14 days (see references in Janetos et al. 2008). Consistent anecdotal observations have been reported in Texas (Middleton 2008).

In an analysis of 866 peer-reviewed papers exploring the ecological consequences of climate change, nearly 60 percent of the 1,598 species studied exhibited shifts in their distribution or phenology over 20–140 years (Parmesan and Yohe 2003). Migration shifts as great as 5.1 days per decade were documented (Root et al. 2003), with an average of 2.3 days per decade across all species (Parmesan and Yohe 2003). In general, the migration of butterflies in spring is highly correlated with spring temperatures and with early springs. Researchers have documented many instances of earlier arrivals and distributional or range shifts in response to warming. Across all studies included in her synthesis, Parmesan (2006) found that 30–75 percent of species had expanded northward, and less than 20 percent had contracted southward. In England, however, no evidence was found for a systematic shift northward across all species of butterflies monitored (Hill et al. 2002), illustrating the need for accurate long-term data to test such hypotheses for specific taxonomic groups and regions.

Pinpointing how the composition of communities will change poses another set of questions (IPCC 2007). Different species will respond differently to scenarios of more or less precipitation. For example, extreme rainfall events can also influence vulnerability of native plant communities to invasion by introduced species such as Chinese tallow (Siemann et al. 2007). The seasonality of rainfall or duration of drought could be critical for some species. For example, would sustained drought limit survival of vulnerable oak species in the Edwards Plateau, and how would this affect those species listed as endangered and threatened that depend on oaks?

Interaction of Habitat Fragmentation and Climate Change

The ability of species to respond to changes in climate is inhibited by reductions in available habitat and increasing isolation of remaining ecological communities resulting from human land-use practices (IPCC 2007). Two factors are important regarding human impact on the landscape: the degree of habitat fragmentation and the availability of corridors of dispersal across gaps between fragments.

Habitat fragmentation refers to the process by which stands of native vegetation become smaller and discontinuous as a result of clearing land for agricultural, residential, and commercial use. The effects of habitat fragmentation vary depending on the requirements of the isolated species, the size and shape of the fragment, existing seed banks for plants, and the abilities of native and nonnative species to move between fragments.

Parks and reserves tend to represent fragments of previously extensive natural vegetation, either because areas with native vegetation were targeted as high priority for protection, or because human development eventually occurs up to the boundaries of protected areas. Fragments of native vegetation may also occur on private lands, depending on the land-use practices of the owners. Biologists refer to the "mosaic" of vegetation types in a given landscape, including those in various stages of cultivation and succession of native vegetation.

Dispersal corridors are relatively long, thin areas connecting habitat fragments; within them, the vegetation structure is more favorable for the movement of native species than in adjacent areas (Saunders and de Rebeira 1991). For example, native vegetation often is allowed to remain along the banks of rivers. If two biological reserves (habitat fragments) are adjacent to the river, then animals and the seeds of plants may move along the river corridor between the two habitat fragments (Packard and Cook 1995). Many factors influence the suitability of strips of native vegetation as corridors, and the behavior of each species must be considered in assessing whether a strip will function as a movement corridor or as a predator trap.

Systems of several reserves connected by corridors have been proposed as more likely to protect biological diversity than isolated protected areas. The logic behind this principle is the same as the old saying, "Never put all your eggs in one basket." If a species becomes extinct in one area, the area may be recolonized from connected habitat fragments (Saunders and de Rebeira 1991). In general, the smaller the fragment, the higher the probability of extinction due to loss of genetic diversity, catastrophe, or an imbalance in relations of competing species, predators, and their prey.

Isolation of a biotic community due to human-related fragmentation may increase the probability of extinction of species as the climate changes. If barriers exist between two fragments of habitat, and one fragment declines in quality, it is unlikely that a species will escape extinction by crossing a barrier to another fragment that is better for it (Peters 1989). This generalization obviously depends on the dispersal abilities of species.

Species Vulnerability

Climate change could cause the disappearance of communities and species from areas where they occur today (Peters 1989). This could happen as a result of several factors: (1) existing microclimatic conditions may shift outside the boundaries of protected areas, (2) species may not be able to migrate as quickly as conditions change, and (3) as other species move into the specific location, new interactions may result. Endemic and rare species are most vulnerable.

Theoretically, if a protected area is at the southern edge of the distribution of an endangered species, that species may disappear from the area as climatic conditions required by the species move northward or toward the tops of mountains (Packard and Cook 1995:325). In an undisturbed system, where habitat fragmentation has not occurred, species that have their center of distribution farther south are likely to invade protected areas to the north as climate warms. Extinction is likely, however, if the means of dispersal do not exist between old, declining fragments and new, improving patches. Furthermore, endangered species adapted to a very specialized, narrow set of conditions are more likely to be adversely affected by climate change than those that have a wide tolerance of climatic conditions but are endangered by human impacts such as overharvesting.

Although endangered species are still viewed by some conservation practitioners as indicators of the health of ecosystems, attention has shifted from single-species management to management of the community of plant and animal species that the endangered species depends on in a given landscape. This management approach raises many questions. Is the current assemblage of species in a community likely to remain constant with climate change and simply shift to a new location? If not, and the species composition of a community shifts, will invading species fulfill the same functions as the previous species that were extirpated? How are resource managers to prepare for the uncertainty?

The rate of habitat change projected under future climate change will be many times greater than the rate at which plant species responded to temperature changes in previous epochs related to glacial periods (Vitousek 1989; Risser 1990). Furthermore, plant and animal species may move at different rates (Davis and Zabinski 1991). Plant species that have windblown seeds will invade new areas more rapidly than those that reproduce vegetatively or whose heavy seeds fall below the parent plant. Likewise, animals that can fly or otherwise migrate long distances between suitable habitat patches will be more resilient than those unable to do so. Some scenarios project movement rates of vegetation zones that are faster (4.5 miles, or 7 kilometers, per year) than rates of range expansion documented for vertebrates such as deer (1–2 miles, or 2–3 kilometers, per year; Peters 1989).

Extinctions can occur as a result of the invasion of antagonistic species that affect the quality of habitat fragments. A warmer climate may favor invasions of insect species previously limited in their northward expansion by the freeze line (the threshold of temperature tolerance for most plants). Such range expansions can affect the viability of native species in the locations invaded. For example, the rapid range expansion of imported fire ants from Alabama to eastern Texas appears to have affected the diversity of ground-dwelling species vulnerable to their attack.

Theoretically, an ideal strategy for protection of a representative sample of biological diversity within a state would include a series of interconnected areas arranged along gradients of expected climate change. In this manner, species could move at their own rate along the gradient. Where this is not possible, restoration technology may aid in artificially moving those species most vulnerable to extinction as threshold conditions shift along the climate gradient (Hoegh-Guldberg et al. 2008).

Complex Interaction of Ecosystem Determinants

Because of the complexity of ecosystems, substantial scientific resources are required to build integrated models that will provide decision makers with the information they need. Nonlinear responses to combinations of key limiting factors for native plants such as precipitation, temperature, soil nutrients, and carbon dioxide in experimental settings suggest that it would be unrealistic to extrapolate from simple models of vegetation response to climate change to entire ecoregions (Gordon and Huxman 2007).

The complex interactions of shrub encroachment, water catchment, and evapotranspiration are under investigation (Brown et al. 2001; Browning et al. 2008), and they vary by species (Wilcox et al. 2006; Patrick et al. 2007). Seasonality of rainfall relative to the growing season of plants can have an important impact. For example, studies of vegetational changes in the Chihuahuan Desert of New Mexico suggest that the vegetation is shifting from grasses to shrubs as a result of changes in the seasonality of drought. Wetter winters favor the growth of shrubs, and drier springs reduce the proliferation of grasses. Shrubs can take advantage of rainfall during the winter and endure the droughts of summer. Grasses, however, do not grow during winter (despite rainfall) and are susceptible to dry conditions in the spring and summer.

For Texas, projections regarding summer rainfall are highly variable. With higher temperatures there will be increased evaporative demand, which could exceed any increase in precipitation (see Chapter 2). This uncertainty also extends to changes in the seasonality of precipitation, which is affected by other factors such as El Niño conditions. Aquatic systems, particularly wetlands, are likely to be most vulnerable to changes in rainfall (Burkett and Kusler 2000).

Better resolution of regional models of climate change will be necessary before it is possible to refine projections of the most likely ecological changes in Texas. Nonetheless, it is possible to carry out sensitivity analyses by looking at an envelope of possible changes in temperature and precipitation and using tools such as ecological niche models. Such experiments have been carried out in Mexico (Peterson et al. 2002) and California (Parra and Monahan 2008) and could be undertaken in Texas. Nevertheless, having made such disclaimers about our predictive abilities given current information, it is possible to identify the salient questions that need to be asked regarding potential changes in each of the natural regions of Texas.

POTENTIAL CHANGES IN EXISTING NATURAL REGIONS OF TEXAS

Vegetation in each natural region of Texas is likely to respond differently to climate change, as determined by regional variation in land use and seasonality of temperature and precipitation (Griffith et al. 2007). Furthermore, the projected influences of climate change on biodiversity in Texas should take into account the cumulative impacts of historical and predicted drivers such as pollution, habitat fragmentation, invasive species, and overexploitation of natural resources (Fig. 5.2). Trends associated with these drivers have been predicted on a national level (Janetos et al. 2008). In this section, we align ecoregions identified on a national and global scale with the biological units used for management of biodiversity in Texas. Such alignment is needed for policymakers in Texas to attract federal support to match priorities identified on a regional basis in the state (addressed in the conclusion of this chapter).

In the following subsections, the climatic characteristics and vegetation of natural regions will be described for Texas, and primary questions regarding potential changes in biodiversity will be identified for each region. We will align the specific terms used for natural regions in Texas with the general terms used on a national and global scale. Our reasoning is that predictions made on a broad scale may or may not be valid for the specific landscape and climatic patterns in Texas. Policies set at a national level will need to be interpreted by practitioners and policymakers at the state and regional levels of governance.

For the purposes of this discussion, we have combined the 12 ecologically distinct vegetation units introduced in Figure 5.1 (as recognized by the

GLOBAL ECOREGION	In Texas	DRIVERS OF BIODIVERSITY CHANGE:				
		Climate change	Pollution	Habitat change	Invasive species	Over-exploitation
Forest	Temperate pine/deciduous	↑	↑?	↘?	↑	→
Dryland	Temperate grassland	↑	↑	↑	→	→
	Tropical grassland & savanna	↑	↑	↗	↑	→
	Desert-Chihuahuan	↑	↑	→	→	→
Mountain	Rocky range to Sierra Madre	↑	↑?	→	→	→
Island	Coastal barrier islands	↑	↑	→?	→	→
Inland	River, lake, wetland, etc.	↑	↑	↑	↑	→?
Coastal	Beach, marsh, bay, estuary	↑	↑	↗	↗	↗
Marine	Coral, fish, shellfish, etc.	↑	↑	↑	→?	↗

	historical impact:	least	predicted impact:		most
				↑	increase
KEY TO		..		↗	increase
IMPACT OF DRIVER		..		→	no change
ON ECOREGION		most		↘	decline

FIGURE 5.2. Factors in biodiversity change. Texas policymakers are urged to consider both the historical (gray intensity) and the projected (direction of arrows) cumulative impacts of multiple drivers likely to interact with climate change and their effect on biodiversity (adapted from Janetos et al. 2008). In the broader ecoregions that are also represented in Texas, federal policy is likely to be guided by the predictions shown here. A statewide interdisciplinary panel is needed to fine-tune these predictions based on regional trends in land use, which may differ from national trends. Also, change is likely to differ at the edges of the national ecoregions, several of which intersect in Texas. Trends where the authors suggest the state perspective is likely to differ from the national perspective are indicated in this figure with bold question marks.

U.S. Environmental Protection Agency; Griffith et al. 2007) into these 7 natural regions widely used for educational purposes by the Texas Parks and Wildlife Department (TPWD): (1) Pineywoods (South Central Plains), (2) Central Texas Savannah (Cross Timbers, Texas Blackland Prairies, and East Central Texas Plains), (3) High and Rolling Plains (High Plains, Southwestern Tablelands, and Central Great Plains), (4) Southern Texas

Plains (subtropical plains and savannah), (5) Chihuahuan Desert (including Arizona–New Mexico mountains that grade into the Sierra Madre in Big Bend), (6) Western Gulf Coastal Plain (coastal prairies, marshes, and islands in southeastern Texas), and (7) Hill Country (Edwards Plateau, unique to Texas).

We have added a subsection for aquatic systems that drain and flood landscapes, thereby connecting these upland systems with the coastal and marine systems that include barrier islands and an offshore coral reef. Both islands and coral reefs have been included in projections of impacts of climate change on biodiversity on a national scale. We include them in this chapter to emphasize opportunities for alignment of state and federal programs addressing effects of climate change on biodiversity in Texas.

Temperate Forests: The Pineywoods of East Texas

Historically, pines and oaks blanketed the low rolling hills of the Pineywoods in East Texas, while tall hardwoods filled the seasonally flooded bottom-lands. The climate in this area is cooler and wetter than other portions of Texas; thus, it better supports the growth of trees. This is the southwestern edge of the southern temperate forests, nationally categorized as the South Central Plains (Fig. 5.1; Griffith et al. 2007).

The original vegetation in much of this region has become fragmented as a result of forest plantations, agriculture, urbanization, and changing land use at the urban interfaces with wildlands (Callicott et al. 2007). Some game species, such as turkey and deer, flourish in a mosaic of vegetation patches that include an understory of diverse species. Other, rare species with specialized requirements, such as the endangered red-cockaded woodpecker, are dependent on fragments of old-growth forest with cavities in trees suitable for nesting. Remnant fragments of bottomland hardwoods along waterways are rich in biological diversity and provide potential travel corridors for wildlife. At the edge of suitable conditions, uncertainty about climate change impacts on this ecoregion in Texas is higher than for southern forests in the southeastern United States (Fig. 5.2).

According to one theory, the boundaries of this forested area would be likely to move toward the northeast (Davis and Zabinski 1991). This would represent an interaction of temperature increase and the projected eastward movement of the zone of soil moisture retention adequate to support an oak-hickory-pine forest in Texas. In addition, the species composition of forest communities might change, because species differ in their response to increases in carbon dioxide levels and water stress. Theoretically, fragments of bottomland hardwoods along rivers would show less change than upland forests, presuming there are no human-related changes in flooding patterns of rivers.

Dryland Temperate Grassland: The Cross Timbers and Savannah of Central Texas

Located west of the Pineywoods is a grassland-forest transition zone, which is included in the broader category of the dryland temperate grassland region. The diversity of vegetation and geological features within this transition zone includes oak woodland, cross timbers, bottomland forests, and prairies. Historically, the growth of trees in this region was probably controlled by fires, drought, and human influences (Albert 2007). The climate in this region is humid subtropical to subtropical subhumid, a little drier than the forests to the east and cooler than the plains to the south (Fig. 5.1).

Natural prairie vegetation has become fragmented where the bands of rich black soils are cultivated in the Cross Timbers. Natural patches of oaks remain interspersed with meadows in regions predominately used as cattle pasture. Strips of oak woodland and prairie run in a southwest-northeast direction, in concert with geologic features and parallel to the predicted gradient of change in soil moisture retention. More information is needed to understand the connectivity of these remaining fragments.

Several alternative hypotheses might be posed regarding future vegetation changes in this region (Fig. 5.2). Those tree species limited by soil moisture retention might be expected to retreat along gradients to the northeast, resulting in a decline of woodlots. Where the expansion of brush species from the south was previously limited by severe freezes, species more typical of the South Texas Plains might be expected to invade. Alternatively, if differences in carbon pathways are important in moderating the response to climate change, shrubs and trees might replace grasses under conditions of carbon dioxide enrichment.

The geological features of Texas are dominated by the southeasterly recession of the coastline and the southeasterly direction of river flows (Packard 1995:133). Thus, the rivers run perpendicular to the bands of geologically uniform terrain and could act as barriers to movement along the predicted gradient of soil moisture retention.

Dryland Temperate Grassland: The Rolling and High Plains of North Texas

Historically, the grasses of the flat plains in the Panhandle marked the southernmost extension of the vast prairies known as the dryland temperate grassland of North America. Ecologists distinguish between two natural regions, the Rolling Plains and the High Plains (Fig. 5.1). The flat, high plains supported immense herds of buffalo and pronghorn antelope and vast prairie-dog towns. In the canyons, cut by the headwaters of rivers draining from the Rolling Plains toward the Gulf, are remnants of the pinyon-juniper-oak biota that once extended to the Rocky Mountains, during a cooler epoch.

Native vegetation in the High Plains has been almost entirely eliminated by agricultural development. Only small fragments of prairie remain, and playa lakes provide important habitat for waterfowl in a mosaic dominated by agricultural lands. A geological break, the Palo Duro Canyon, represents a unique geographical feature. It is difficult to predict how the canyon's microclimate would respond to global warming, with the combined impacts of grazing on native plants by exotic and native herbivores and the stresses of climate change.

Without additional rainfall, wetlands associated with playa lakes might shrink in size and dry up more quickly on a seasonal basis. Changing land use from rangeland to irrigated agriculture in this region has a tremendous influence on biodiversity in the High Plains (Scanlon et al. 2005), so cumulative impacts should be considered in assessing projected impacts of climate change (Fig. 5.2).

Subtropical Grassland and Savannah: Brushlands of South Texas

The South Texas brushlands are at the northern edge of the subtropical grassland and savannah, which extends almost to the Tropic of Cancer. This subtropical, subhumid desert jungle is unique to Texas and Mexico. Influenced by storms from the Gulf, it nevertheless receives sporadic rainfall and is susceptible to droughts. This semiarid climate is characterized by warm Neotropical temperatures, and the humidity is higher than in the arid, higher-elevation western portion of the state.

Distinctive differences in plant communities exist within this region along rivers, on the flat plains, and along the coast. The Coastal Sand Plains, typified by the expansive lands of the King Ranch, are well known as good wildlife habitat and extensive cattle range. Much of this region is excellent habitat for deer and is easily managed for quail hunts (Brennan 2007). Denser brush along drainages provides good habitat for javelina, also a game species. The native pecan woods that used to flank rivers have been reduced to small fragments, but they still provide important habitat for diverse species of birds.

It is difficult to predict how animals will adapt to climate change in South Texas; some will do better than others (Brennan 2007). South Texas provides wintering habitat for several dozen short-distance migrants, and climate changes may result in a reduction in the amount of time they spend there. With respect to northern bobwhites, a 4°C (7.2°F) increase in ambient temperature could have a deleterious effect on productivity by reducing the number of egg-laying days (Brennan 2007).

Most of the brush has been cleared from the rich Rio Grande Valley, the delta, and the banks of the Lower Rio Grande. Vegetation in this subtropical region appears to be important as a staging area for Neotropical migratory songbirds, and a wide diversity of bird species can be sighted in the region because of the overlap in ranges of bird species from temperate and tropical regions. Specific locations, such as the shallow waters in the salt pans of Sal del Rey, attract a high diversity of shorebirds.

Theoretically, a change in temperature would have little effect on the plant communities in the brushlands region, since most are adapted to the warm climate (Fig. 5.2). We do not know, however, whether sites dependent on shallow water, such as Sal del Rey, would be adversely affected because of greater rates of evaporation.

Dryland Desert and Mountains: Chihuahuan Desert of West Texas

The wide-open panoramas, short brush, arid hills, and mountains of West Texas are classified as dryland desert and mountains. Warm and dry climatic conditions are typical of the Chihuahuan Desert on this high-elevation plateau (Fig. 5.1). Species typical of the mountains in Arizona and New Mexico interface with species at the northern extent of their range in the Sierra Madre in the Trans-Pecos and Big Bend region of Texas. For example, nectar-feeding *Leptonycteris* bats migrate seasonally along these mountain ranges following the phenology of flowering agave, distributed in patches popularly described as "mountain islands in desert seas."

The Chihuahuan Desert is influenced by humid weather systems from the Gulf of Mexico during the summer, and during the winter it receives precipitation transported sporadically by storms in the Pacific. Since the growth of woody plants is a complex interaction of precipitation and evapotranspiration (Scott et al. 2006), some researchers project that woody brush encroachment in this region could influence stream flows and springs (Huxman et al. 2005). Changes in perennial grass production in this region have important implications for Chihuahuan Desert systems in Southwest Texas (Khumalo and Holechek 2005). An increase in summer precipitation would likely increase carbon sequestration in the sotol grasslands of Big Bend (Patrick et al. 2007).

The mountains and scattered hills in this desert region provide much variation in terrain and vegetation. The taller Davis and Guadalupe mountains in the northern portion of this region are most closely affiliated with the southern end of the Rocky Mountain range, whereas the biota of the Chisos Mountains in the south are most closely associated with the northern tip of the Sierra del Carmen range, the first wilderness area in Mexico to be protected by joint private and public partnerships.

Two rivers in this region represent a distinctive habitat type that contributes to the diversity of the biota. The Rio Grande cuts a west-east transect, and the Pecos River has a southeastern course to its junction with the Rio Grande. The seasonal flow of the Rio Grande has been compromised not only by land and water use changes in its watershed but also by changes in the mountains of the Sierra Madre that feed its tributaries, such as the Rio Conchos from Mexico.

Theoretically, the greatest changes in vegetation likely to occur with warming of this region are movements of vegetation zones upward in altitude. Rare plants adapted to medium altitudes may undergo habitat fragmentation as suitable conditions disappear from the valleys and are limited to isolated hills.

Gulf Coast Plains and Saltwater Systems: Marshes, Estuaries, and Barrier Islands

The biodiversity of the coastal plains of Texas is influenced by barrier islands stretching along the coast, salt-grass marshes surrounding bays and estuaries, and inland prairies and bottomland hardwoods flanking some rivers. The shoreline runs parallel to the gradient of rainfall, and climatic variation is primarily influenced by storms from the Gulf. Seven rivers empty into bays and estuaries along the coast of Texas, each influenced by distinct combinations of conditions due to the hydrology of the watersheds, urban and agricultural water demands, and vegetation.

The Gulf Coast region is widely known for its importance as an overwintering region for shorebirds and waterfowl migrating from the Canadian Arctic along the Central Flyway (Reid and Trexler 1991). In addition to coastal marshes, barrier islands and estuaries are important for shorebirds. Ducks move between feeding areas in the estuaries and inland freshwater ponds. Geese feed and loaf in the mosaic of salt marshes, rice fields, and inland wetlands. The population of the endangered whooping crane has been stabilized with protection of the winter feeding grounds on shallow mudflats and islands, but changes in water salinity associated with changing river flows could threaten the abundance and distribution of blue crabs, the cranes' primary food source. Areas currently protected may no longer be sufficient when sea level rises.

Even for migratory species the picture is complicated (Kim et al. 2008). With global warming, vegetation changes are projected to be greater near the poles than near the equator. Thus, loss of tundra nesting habitat may threaten the migratory shorebirds, ducks, cranes, and geese that winter in Texas (Peters 1989). Warmer winters may favor subtropical species that migrate north in search of more food supplies in southern reaches of the coastal plains.

The projected effects of climate change on increasing salinity of estuaries and flooding of salt marshes are discussed in more detail in Chapter 4. In general, coastal systems are stressed by multiple changes and are not likely to be resilient to the additional consequences of climate change. The Union of Concerned Scientists and the Ecological Society of America analyzed climate change impacts on the ecological heritage of the Gulf Coast region in a 2001 report (Twilley et al. 2001).

Edwards Plateau: Unique to the Hill Country of West Central Texas

The scenic Edwards Plateau is well known for its savanna, a grassland interspersed with oak-juniper woodlands, that grows on dissected, karstic limestone (Fig. 5.1). The climate, slightly cooler than the south and drier than the east, is known as subtropical steppe. The geology is distinctly different from other regions because of the ancient granitic uplift of overlaying limestone deposited during an epoch when this part of Texas was covered by an inland sea.

Thirty-three species listed as endangered or threatened are found only in the Edwards Plateau (TPWD 2003; Campbell 2003). The limestone karst in this region has numerous caves, sinkholes, and fissures where water has dissolved the limestone. This karst provides habitat for unique invertebrates, including 19 that are endangered (U.S. Fish and Wildlife Service 2008). The springs flowing from the underground aquifer provide specialized habitat for one plant, four fish, and several salamander species listed as threatened or endangered by the TPWD (2003). The quality and quantity of water flow from surface and groundwater sources are currently adversely affected by multiple factors, including urbanization (U.S. Fish and Wildlife Service 2008). Potential interactive effects among climate change, urbanization, and municipal and agricultural water demands are likely to be complex in the Edwards Plateau.

The complexity of this landscape is further influenced by specific geography in the Edwards Plateau (Wilcox et al. 2007), year-to-year variation in the way shrubs utilize water sources (McCole and Stern 2007), and microbial biomass (McCulley et al. 2007). For many native species of plants in this area, the extremes of temperature and rainfall have a strong influence on survival. Stress due to climatic factors near the threshold of tolerance may make individual species more susceptible to diseases or other pathogens. Each species has a unique set of tolerance ranges to physical factors, and such tolerances change during the lifetime of the individual, further complicating projections of biodiversity changes.

Aquatic Systems: Freshwater, Brackish, and Saltwater

In Texas, a rise in sea level could result in loss of coastal wetlands, erosion of beaches, and intrusion of salt water into groundwater supplies, as detailed in Chapter 4. Interactive effects of multiple stressors need to be considered for aquatic systems. The salt tolerance of some species could be exceeded, causing changes to food webs and possibly a reduction in biological productivity and diversity. Evaporation rates are likely to increase as the climate warms, which could result in decreased river flows, drops in lake levels, reduced groundwater recharge, and diminished freshwater inflows to bays and estuaries. Increased rates of stream flow in some areas, driven by increases in precipitation or other mechanisms, could offset some of those effects. Higher water temperatures are likely to alter freshwater species assemblages through reductions in water quality, particularly via reduced levels of dissolved oxygen.

Increases in storm surges and a rise in sea level are projected through the twenty-first century (Hopkinson et al. 2008). The extent to which these changes will displace the wetlands in coastal plain communities of Texas is unknown and needs to be modeled. Changes in stream flow may also impact estuaries, and projected changes are beginning to be examined (Forbes and Dunton 2006; Makkeasorn et al. 2008; Russell and Montagna 2007; Tolan 2007). Sand-dune communities on barrier islands and along the coast may be particularly vulnerable to sea-level rise (Feagin et al. 2005).

The geographic ranges and limits of many aquatic and wetland species are defined by temperature (Poff et al. 2002). Some species may be locally extirpated if the water temperature exceeds their thermal tolerance. For example, larval production of the endangered fountain darter decreases significantly at 77°F (25°C) and above (McDonald et al. 2007), and significant lethality occurs above 89°F (32°C) (Bonner et al. 1998).

Warm waters are naturally more productive than cool waters, but nuisance algal blooms are expected to increase as temperature continues to rise. With precipitation projections highly uncertain, it is difficult to ascertain to what degree there may be synergistic effects of climate change on water quality. For example, reductions in summertime runoff and elevated temperature would likely exacerbate any existing impairments of water quality, with concomitant impacts on aquatic biota.

Finally, invasive species, which tolerate a range of thermal conditions different from that of native species, may mediate novel interactions resulting in compromised aquatic systems. For instance, the introduced giant ramshorn snail poses a threat to the threatened and endangered species associated with the San Marcos and Comal rivers, because it grazes on aquatic plants and can withstand temperatures to 102°F (39°C) (McKinney and Sharp 1995).

BIODIVERSITY, CLIMATE CHANGE, AND DECISION MAKING

What are the implications of potential changes in native ecosystems for the decision makers of today? The greatest challenge and opportunity facing Texas and its resource agencies is the predominance of private landholdings, by some estimates in excess of 95 percent of the state. Compounding the challenge, Texan decision makers have relatively few data on long-term trends for species, making it difficult both to assess responses to climate change and, more fundamentally, to understand what resources may be at risk.

The mission of the Texas Parks and Wildlife Department encompasses the protection, conservation, and management of biodiversity. The department has several documents that guide biodiversity conservation, including the *Land and Water Resources Conservation and Recreation Plan* (TPWD 2010) and the *Texas Comprehensive Wildlife Conservation Strategy* (TPWD 2005), also known as the state's wildlife action plan. The land and water plan mentions climate change as an emerging issue that needs to be addressed. Likewise, the 2010 update of the wildlife action plan will also address climate change. The TPWD houses the Texas Natural Diversity Database, which serves as a clearinghouse for biodiversity data and is a member organization of NatureServe.

Texas state agencies are well positioned to learn from other joint efforts. For example, conservation planning efforts elsewhere have combined genetic and ecological data to evaluate the "multi-species genetic landscape" (Vandergast et al. 2008; Davis et al. 2008). The Climate Impacts Group at the University of Washington has demonstrated the value of an ecosystem approach, linking adjoining states in the analysis of climate impacts on the Pacific Northwest (Ruckelshaus and McClure 2007).

General Guidelines

The sites most vulnerable to global climate change will be those that are smallest and most isolated from other natural areas (Hopkins et al. 2007), for example, remnants of prairie vegetation in the High Plains and Rolling Plains. Areas that are connected by patterns of land use that encourage movement of native species between fragments of habitat may be less vulnerable.

The protected species most vulnerable to extinction will be those that are adapted to specific microclimates likely to disappear with global warming and those that are unable to move long distances between suitable habitat fragments (Hopkins et al. 2007), for example, karst invertebrates in the Balcones Canyonlands or populations isolated on mountains in West Texas. As populations of protected species become isolated in small fragments of habitat, they are more likely to become extinct (the plight, for example, faced by Attwater's prairie chicken). Alternatively, those that live together in large areas transfigured by human land uses may become more susceptible to the combined stresses that they encounter as a result of climate change, such as saltwater intrusion into marshes of the Gulf Coast plains between Houston and Beaumont.

Beyond these general guidelines (Hopkins et al. 2007), it is difficult at this time to predict what changes will occur, because each ecosystem and species has its unique set of adaptations and interrelations with the physical world. One factor that decision makers should examine carefully is whether the basic structure of protected areas will facilitate the transition to a changing climate.

Regional Mosaics of Protected Areas

The analysis required to determine whether there is sufficient habitat protection to sustain viable populations of native species is complex and requires a regional approach (Scott et al. 1991). This type of analysis is known from model programs in other states as GAP analysis. It goes beyond identifying areas protected for scenic and recreational values by examining the utility of landscapes for maintaining sets of native species with complementary habitat requirements.

An example of this regional approach is the Balcones Canyonlands National Wildlife Refuge, established to provide habitat for two endangered species, the black-capped vireo and the golden-cheeked warbler. An adaptive management approach with systematic monitoring is needed to determine the extent to which representative fragments of habitat currently protected on private and public lands are sufficient for population viability. An additional question is whether there is sufficient connectivity between protected areas within the region, as required for dispersal to occur as climatic conditions shift.

Endangered Species

Rising human population density is putting many ecosystems at risk. Texas has lost most of its prairies and almost all of its native habitat in the lower delta of the Rio Grande and in regions surrounding metropolitan areas. Texas has 176 state-listed and 93 federally listed threatened and endangered species and 20 federal candidate species (TPWD, personal communication).

Texas has separate laws to protect plants and animals on the state list (Texas Parks and Wildlife Code Ann. sections 68.001 et seq.; 83.006; 49.015, Texas Admin. Code title 30, sec 330.129). For animals, listings are based on scientific data only. The law does not require recovery plans, critical habitat designation, or agency consultation, although by regulation a preliminary investigation of land to be developed is required. The law covering plants provides for listings (Texas Parks and Wildlife Code Ann. sections 88.001 et seq.). In addition, listed plants cannot be collected from public lands without a TPWD permit and cannot be collected from private lands for commercial sale without landowner permission and a TPWD permit analysis (NHNCT 2008).

Protection of Habitat on Private Lands

Texas offers several private land conservation programs with the dual benefits of protecting water and wildlife resources (Wagner et al. 2007). For example, conservation easements are authorized by statute (Texas Natural Resources Code Ann. sections 183.001 et seq.). More than 22 million private acres are under active, written TPWD wildlife management plans that encourage public-private partnerships for biodiversity conservation. In addition, a 1995 constitutional amendment allows open space used for wildlife management to qualify for tax abatement in the same manner as open space used for agriculture (Texas Constitution Article 8 section 1-d-1). The State Wetlands Conservation Plan promotes conservation through incentives for private landowners.

Successful public-private partnerships utilizing a diverse set of conservation tools for land stewardship have been established throughout the continent (Groves 2003) and are stimulating innovative approaches adapted to the sociopolitical climate of Texas (National Park Service 2008). For example, the Pineywoods Mitigation Bank has been established for restoration and protection of bottomland hardwood habitat in river floodplains (Conservation Fund 2008). The Texas Land Trust Council, an umbrella organization, reports approximately 570,000 acres of land under conservation easements in Texas, with even more acreage protected by fee simple agreements. Innovative approaches have been used successfully by municipalities and conservation organizations throughout the state.

Restoration and Ecosystem Management

Restoration science is an expanding field, as resource managers recognize that most ecosystems bear the imprint of human activity and are substantially altered from their historic state. Where there had been resistance among some ecologists to the concept of tampering with nature, a growing chorus supports the idea of actively managing these novel ecosystems (Seastedt et al. 2008). For instance, wetland restoration is now an accepted technique in mitigation proceedings.

Assisted colonization, the idea of deliberately moving species, is being advocated by some ecologists as a tool to aid species that might otherwise be unable to disperse or adapt fast enough to climate change (Hoegh-Guldberg et al. 2008). In the past, the strategy of intensive management, requiring relocation of individuals to maintain genetic diversity in small populations, has been pursued where habitat fragments became too isolated for natural recolonization. Assisted colonization will need to be supported by sufficient human, technological, and economic resources if it is to succeed. It may offer viable options in the face of potential conflicts among private, local government, state, and federal interests.

Policy Alternatives

Governments have been encouraged to take anticipatory rather than reactive measures in adapting to climate change and to address the regulations protecting endangered species (Powledge 2008). Once a species is extinct, reactive measures will not revive it. In general, proactive measures are less costly than restoration.

In evaluating alternative options for anticipatory actions, Smith et al. (1991) recommend using the criteria of flexibility, economic efficiency, feasibility, and consideration of associated benefits. They identify several policy options with regard to natural systems: (1) strengthen and enlarge existing protected areas, (2) establish migration pathways between existing protected areas, (3) protect areas that may become suitable habitat for threatened and endangered species in the future, (4) increase restrictions on zoning and management around reserves, (5) avoid permanent alterations to rivers and streams, which may be important migratory pathways under changing climate, (6) evaluate species stocking and introduction strategies, (7) reduce destruction and pollution of habitats in general, and (8) adjust species preservation programs to more broadly protect habitat and ecosystems. A similar set of guidelines published in England recommends integrating mitigation and adaptation measures into management, planning, and practice to meet biodiversity conservation goals (Hopkins et al. 2007).

These policies are all very appropriate from the biological standpoint and can aid adaptation to climate change. They may need to be modified for specific regions and sociopolitical groups. For example, in some groups, zoning restrictions around reserves are seen as an infringement on private property rights and abuse of government power. Most of the land in Texas is under private ownership; therefore, viable populations of native species are not likely to be maintained solely on public lands. Workable, mutually beneficial solutions need to be further developed to compensate private landowners who protect native species. Given a tight economic climate, a thorough review of existing tax incentives and implementation of additional incentives would be very sound policy.

CONCLUSIONS AND RECOMMENDATIONS

Texas boasts a rich heritage of biodiversity in terms of the variety of ecoregions, species, and genotypes occurring within its unique geographic setting. Climate change represents just one of a set of stressors challenging fauna and flora, whose resilience is already at risk from land development, habitat fragmentation, invasive species, chemical stressors, and direct exploitation (Table 5.1). Across the nation, decision makers are being tasked to develop, without sufficient information, policies to protect biodiversity in the face of climate change (Powledge 2008).

What information would help resource managers confront the challenges posed to biodiversity by climate change? A necessary first step is the targeted assessment of fish and wildlife species and habitats in order to identify particularly vulnerable populations. If key areas are protected, they could provide migration corridors or replacement habitat. Habitat restoration might enhance carbon sequestration while also helping species cope with climate change. Each natural region needs to be examined in detail to determine the likely shifts in distribution of native species within the existing mosaic of agricultural, urban, and natural areas (Fig. 5.2). Some unique, isolated features, like the hotspot of species diversity in the Edwards Plateau, are likely to be the least resilient and require special attention.

From a research perspective, projections of land-use and land-cover changes would be coupled with projected climate changes in spatially explicit habitat models linked to individual-based plant and animal population models in order to project population trends (McRae et al. 2008). Such a modeling framework holds the promise of allowing managers to begin exploring the complex interactions that may not be apparent from simple modeling approaches (e.g., isolated populations of endangered species). This multi-model framework would require analysis of the interactions of natural and human systems within specified envelopes of climate change. This is the path California took when it initiated the Scenarios Project in 2005 and invested substantial resources in exploring the range of possible climate futures for the state and the impact those futures would have on six sectors, including the natural environment (Cayan et al. 2008).

A proactive approach to managing the natural heritage of Texas in the context of climate change will be more cost-effective than reactive measures. Texas possesses one of the richest natural heritages in North America because of its location at the continental intersection of forest and desert biomes and temperate and subtropic climates (Fig. 5.1). The effects of climate change are likely to be more pronounced at edges like this. Considerable effort has been invested in protecting this living library, valued by local and global communities. The quality of this investment needs to be protected into the future, from both legal and ethical perspectives.

Regardless of the exact approach undertaken, city, regional, and state networks will need resources to coordinate their efforts with supporting national programs. Expansion of the existing collaborative efforts of private, municipal, state, and federal partners engaged in wildlife management and other conservation programs in Texas will form a strong foundation to protect biodiversity for use and enjoyment by present and future generations.

In the long term, less resilient ecosystems may be unable to provide the ecosystem services on which human societies depend (e.g., clean water, nutrient cycling, pollination, and carbon sequestration). The procurement of new data and implementation of new initiatives come with a cost, but the cost of not acting now could be very, very high.

ACKNOWLEDGMENTS

The authors are very grateful to all who provided invaluable input at various stages in the writing of this chapter.

REFERENCES

Acevedo, M. F., J. B. Callicott, M. Monticino, D. Lyons, J. Palomino, J. Rosales, L. Delgado, M. Ablan, J. Davila, G. Tonella, H. Ramirez, and E. Vilanova, 2008. Models of Natural and Human Dynamics in Forest Landscapes: Cross-site and Cross-cultural Synthesis. *Geoforum* 39:846–866.

Albert, B. M., 2007. Climate, Fire, and Land-use History in the Oak-pine-hickory Forests of Northeast Texas during the Past 3500 Years. *Castanea* 72:82–91.

Archer, S., 1990. Development and Stability of Grass/woody Mosaics in a Subtropical Savanna Parkland, Texas, USA. *Journal of Biogeography* 17:453–462.

Barnard, P., and W. Thuiller, 2008. Introduction. Global Change and Biodiversity: Future Challenges. *Biology Letters* 5. Online DOI: 10.1098/rsbl.2008.0374.

Bonner, T. H., T. M. Brandt, J. N. Fries, and B. G. Whiteside, 1998. Effects of Temperature on Egg Production and Early Life Stages of the Fountain Darter. *Transactions of the American Fisheries Society* 127:971–978.

Brennan, L., 2007. South Texas Climate 2100: Potential Ecological and Wildlife Impacts. In: *The Changing Climate of South Texas, 1900–2100: Problems and Prospects, Impacts and Implications.* J. Norwine and K. John (eds.). Texas A&M University, Kingsville.

Brown, J. H., G. Thomas, S. K. Whitham, E. Morgan, and C. A. Gehring, 2001. Complex Species Interactions and the Dynamics of Ecological Systems: Long-term Experiments. *Science* 293:643–649.

Browning, D. M., S. R. Archer, G. P. Asner, M. P. McClaran, and C. A. Wessman, 2008. Woody Plants in Grasslands: Post-encroachment Stand Dynamics. *Ecological Applications* 18:928–944.

Burkett, V., and J. Kusler, 2000. Climate Change: Potential Impacts and Interactions in Wetlands of the United States. *Journal of the American Water Resources Association* 36:313–320.

Burkett, V. R., D. A. Wilcox, R. Stottlemyer, W. Barrow, W. D. Fagre, J. Baron, J. Price, J. L. Nielsen, C. D. Allen, D. L. Peterson, G. Ruggerone, and T. Doyle, 2005. Nonlinear Dynamics in Ecosystem Response to Climatic Change: Case Studies and Policy Implications. *Ecological Complexity* 2:357–394.

Callicott, J. B., R. Rozzi, L. Delgado, M. Monticino, M. Acevedo, and P. Harcombe, 2007. Biocomplexity and Conservation of Biodiversity Hotspots: Three Case Studies from the Americas. *Philosophical Transactions of the Royal Society, B, Biological Sciences* 362:321–333.

Cayan, D. R., A. L. Luers, G. Franco, M. Hanemann, B. Croes, and E. Vine, 2008. Overview of the California Climate Change Scenarios Project. *Climatic Change* 87(Suppl 1):S1–S6.

Conservation Fund, 2008. *Pineywoods Mitigation Bank.* The Conservation Fund, Austin, Tex. http://www.pineywoodsbank.com/, accessed March 2010.

Davis, M. B., and C. Zabinski, 1991. Changes in Geographical Range Resulting from Greenhouse Warming: Effects on Biodiversity in Forests. In: *Consequences of the Greenhouse Effect for Biological Diversity,* R. L. Peters (ed.). Yale University Press, New Haven, Conn.

Davis, E. B., M. S. Koo, C. Conroy, J. L. Patton, and C. Moritz, 2008. The California Hotspots Project: Identifying Regions of Rapid Diversification of Mammals. *Molecular Ecology* 17:120–138.

Feagin, R. A., D. J. Sherman, and W. E. Grant, 2005. Coastal Erosion, Global Sea-level Rise, and the Loss of Sand Dune Plant Habitats. *Frontiers in Ecology and the Environment* 3:359–364.

Forbes, M. G., and K. H. Dunton, 2006. Response of a Subtropical Estuarine Marsh to Local Climatic Change in the Southwestern Gulf of Mexico. *Estuaries and Coasts* 29:1242–1254.

Fox, D., 2007. When Worlds Collide. *Conservation* 8:28–34.

Gordon, W. S., and T. E. Huxman, 2007. Ecohydrology and Climate Change. In: *Hydroecology and Ecohydrology: Past, Present, and Future.* P. J. Wood, D. M. Hannah, and J. P. Sadler (eds.). John Wiley and Sons, Somerset, N.J.

Griffith, G., S. Bryce, J. Omernik, and A. Rogers, 2007. *Ecoregions of Texas.* Project report to Texas Commission on Environmental Quality, Austin.

Groves, C. R., 2003. *Drafting a Conservation Blueprint: A Practitioner's Guide to Planning for Biodiversity.* Island Press, Washington, D.C.

Hill J. K., C. D. Thomas, R. Fox, M. G. Telfer, S. G. Willis, J. Asher, and B. Huntley, 2002. Responses of Butterflies to Twentieth Century Climate Warming: Implications for Future Ranges. *Proceedings of the Royal Society of London, B, Biological Sciences* 269:2163–2171.

Hoegh-Guldberg, O., L. Hughes, S. McIntyre, D. B. Lindenmayer, C. Parmesan, H. P. Possingham, and C. D. Thomas, 2008. Assisted Colonization and Rapid Climate Change. *Science* 321:345–346.

Hopkins, J. J., H. M. Allison, C. A. Walmsley, M. Gaywood, and G. Thurgate, 2007. *Conserving Biodiversity in a Changing Climate: Guidance on Building Capacity to Adapt.* Department for Environment, Food, and Rural Affairs, London, U.K.

Hopkinson, C. S., A. E. Lugo, M. Alber, A. P. Covich, and S. J. Van Bloem, 2008. Forecasting Effects of Sea-level Rise and Windstorms on Coastal and Inland Ecosystems. *Frontiers in Ecology and the Environment* 6:255–263.

Huxman, T. E., B. P. Wilcox, D. D. Breshears, R. L. Scott, K. A. Snyder, E. E. Small, K. Hultine, W. T. Pockman, and R. B. Jackson, 2005. Ecohydrological Implications of Woody Plant Encroachment. *Ecology* 86:308–319.

IPCC, 2007. Summary for Policymakers. In: *Climate Change 2007: Impacts, Adaptation, and Vulnerability.* Contribution of Working Group 2 to the Fourth Assessment Report of the Intergovernmental Panel on Climate Change. Cambridge University Press, Cambridge, U.K., and New York.

Janetos, A. C., L. Hansen, D. Inouye, B. P. Kelly, L. Meyerson, W. Peterson, and R. Shaw, 2008. Biodiversity. In: *The Effects of Climate Change on Agriculture, Land Resources, Water Resources, and Biodiversity.* P. Backlund, A. C. Janetos, and D. Schimel. Climate Change Science Program Synthesis and Assessment Product 4.3, Washington, D.C.

Khumalo, G., and J. Holechek, 2005. Relationships between Chihuahuan Desert Perennial Grass Production and Precipitation. *Rangeland Ecology and Management* 58:239–246.

Kim, D. H., R. D. Slack, and F. Chavez-Ramirez, 2008. Impacts of El Nino–Southern Oscillation Events on the Distribution of Wintering Raptors. *Journal of Wildlife Management* 72:231–239.

MacMynowski, D. P., and T. L. Root, 2007. Climate and the Complexity of Migratory Phenology: Sexes, Migratory Distance, and Arrival Distributions. *International Journal of Biometeorology* 51:361–373.

Makkeasorn, A., N. B. Chang, and X. Zhou, 2008. Short-term Streamflow Forecasting with Global Climate Change Implications: A Comparative Study between Genetic Programming and Neural Network Models. *Journal of Hydrology* 352:336–354.

McCole, A. A., and L. A. Stern, 2007. Seasonal Water Use Patterns of *Juniperus ashei* on the Edwards Plateau, Texas, Based on Stable Isotopes in Water. *Journal of Hydrology* 342:238–248.

McCulley, R. L., T. W. Boutton, and S. R. Archer, 2007. Soil Respiration in a Subtropical Savanna Parkland: Response to Water Additions. *Soil Science Society of America Journal* 71:820–828.

McDonald, D. L., T. H. Bonner, E. L. Oborny, and T. M. Brandt, 2007. Effects of Fluctuating Temperatures and Gill Parasites on Reproduction of the Fountain Darter, *Etheostoma fonticola*. *Journal of Freshwater Ecology* 22:311–318.

McKinney, D. C., and J. M. Sharp, 1995. Springflow Augmentation of Comal Springs and San Marcos Springs, Texas: Phase I Feasibility Study. Technical Report CRWR 247. Center for Research in Water Resources, University of Texas, Austin.

McRae, B. H., N. H. Schumaker, R. B. McKane, R. T. Busing, A. M. Solomon, and C. A. Burdick, 2008. A Multi-model Framework for Simulating Wildlife Population Response to Land-use and Climate Change. *Ecological Modelling* 219:77–91.

Middleton, R., 2008. The Big Shift. *Texas Parks and Wildlife Magazine* (May).

NHNCT, 2008. Texas Biodiversity Laws and Regulations. A Natural History of North Central Texas website, http://www.nhnct.org/urban/biodivlaws.html, accessed November 2008.

National Park Service, 2008. *Protecting Open Space: Tools and Techniques*. Rivers, Trails, and Conservation Assistance, National Park Service, Austin, Tex.

Packard, J. M., 1995. Biodiversity. In: *The Impact of Global Warming on Texas: A Report of the Task Force on Climate Change in Texas*. G. R. North, J. Schmandt, and J. Clarkson (eds.). University of Texas Press, Austin, Texas.

Packard, J. M., and T. L. Cook, 1995. Effects of Climate Change on Biodiversity and Landscape Linkages in Texas. In: *The Changing Climate of Texas: Predictability and Implications for the Future*. J. Norwine, J. R. Giardino, G. R. North, and J. B. Valdes (eds.). Cartographics, Texas A&M University, College Station.

Parmesan, C., 2006. Ecological and Evolutionary Responses to Recent Climate Change. *Annual Review of Ecology, Evolution, and Systematics* 37:637–669.

Parmesan, C., and G. Yohe, 2003. A Globally Coherent Fingerprint of Climate Change Impacts across Natural Systems. *Nature* 421:37–42.

Parra, J. L., and W. B. Monahan, 2008. Variability in 20th Century Climate Change Reconstructions and Its Consequences for Predicting Geographic Responses of California Mammals. *Global Change Biology* 14:2215–2231.

Patrick, L., J. Cable, D. Potts, D. Ignace, G. Barron-Gafford, A. Griffith, H. Alpert, N. Van Gestel, T. Robertson, T. E. Huxman, J. Zak, M. E. Loik, and D. Tissue, 2007. Effects of an Increase in Summer Precipitation on Leaf, Soil, and Ecosystem Fluxes of CO_2 and H_2O in a Sotol Grassland in Big Bend National Park, Texas. *Oecologia* 151:704–718.

Peters, R. L., 1989. Threats to Biological Diversity as the Earth Warms. In: *Global Change and Our Common Future: Papers from a Forum.* R. S. DeFries and T. F. Malone (eds.). National Academy Press, Washington, D.C.

Peterson, A. T., M. A. Ortega-Huerta, J. Bartley, V. Sánchez-Cordero, J. Soberón, R. H. Buddemeier, and D. R. B. Stockwell, 2002. Future Projections for Mexican Faunas under Global Climate Change Scenarios. *Nature* 416:626–629.

Poff, L., M. Brinson, and J. Day, 2002. *Aquatic Ecosystems and Global Climate Change: Potential Impacts to Inland Freshwater and Coastal Wetland Ecosystems in the United States.* Pew Center on Global Climate Change, Arlington, Va.

Powledge, F., 2008. Climate Change and Public Lands. *Bioscience* 58:912–918.

Reid, W. V., and M. C. Trexler, 1991. *Drowning the National Heritage: Climate Change and U.S. Coastal Biodiversity.* World Resources Institute, Washington, D.C.

Risser, P., 1990. Impacts of Climate Change and Variability on Ecological Systems. In: *The Rio Grande Basin: Global Climate Change Scenarios,* W. Stone, M. Minnis, and E. Trotter (eds.). Water Resource Research Institute, Albuquerque, N.M.

Root T. L., J. T. Price, K. R. Hall, S. H. Schneider, C. Rosenzweig, and J. A. Pounds, 2003. Fingerprints of Global Warming on Wild Animals and Plants. *Nature* 421:57–60.

Ruckelshaus, M. H., and M. M. McClure (eds.), 2007. *Sound Science: Synthesizing Ecological and Socioeconomic Information about the Puget Sound Ecosystem.* Prepared in Cooperation with the Sound Science Collaborative Team. U.S. Department of Commerce, National Oceanic and Atmospheric Administration (NMFS), Northwest Fisheries Science Center, Seattle, Wa.

Russell, M. J., and P. A. Montagna, 2007. Spatial and Temporal Variability and Drivers of Net Ecosystem Metabolism in Western Gulf of Mexico Estuaries. *Estuaries and Coasts* 30:137–153.

Saunders, D., and C. de Rebeira, 1991. Values of Corridors to Avian Populations in a Fragmented Landscape. In: *The Role of Corridors in Nature Conservation.* D. Saunders and R. Hobbs (eds.). Surrey Beatty and Sons, Chipping Norton, Australia.

Scanlon, B. R., R. C. Reedy, D. A. Stonestrom, D. E. Prudic, and K. F. Dennehy, 2005. Impact of Land Use and Land Cover Change on Groundwater Recharge and Quality in the Southwestern U.S. *Global Change Biology* 11:1577–1593.

Schlesinger, W. H., J. F. Reynolds, G. L. Cunninham, L. F. Huenneke, W. M. Jarrell, R. A. Virginia, and W. G. Whitford, 1990. Biological Feedbacks in Global Desertification. *Science* 247:1043–1048.

Scott, J. M., B. Csuti, and S. Caicco, 1991. Gap Analysis: Assessing Protection Needs. In: *Landscape Linkages and Biodiversity.* W. E. Hudson (ed.). Island Press, Washington, D.C.

Scott, R. L., T. E. Huxman, W. L. Cable, and W. E. Emmerich, 2006. Partitioning of Evapotranspiration and Its Relation to Carbon Dioxide Exchange in a Chihuahuan Desert Shrubland. *Hydrological Processes* 20:3227–3243.

Seastedt, T. R., R. J. Hobbs, and K. N. Suding, 2008. Management of Novel Ecosystems: Are Novel Approaches Required? *Frontiers in Ecology and the Environment* 6:547–553.

Siemann, E., W. E. Rogers, and J. B. Grace, 2007. Effects of Nutrient Loading and Extreme Rainfall Events on Coastal Tallgrass Prairies: Invasion Intensity, Vegetation Responses, and Carbon and Nitrogen Distribution. *Global Change Biology* 13:2184–2192.

Siikamaki, J. V., 2008. Biodiversity: What It Means, How It Works, and What the Current Issues Are. *Resources Magazine* (spring). Resources for the Future, Washington, D.C.

Smith, J., A. Silbiger, R. Benioff, J. Titus, D. Hinckley, and L. Kalkstein, 1991. *Adapting to Climate Change: What Governments Can Do.* U.S. Environmental Protection Agency, Climate Change Adaptation Branch, Washington, D.C.

Stein, B. A. 2002. States of the Union: Ranking America's Biodiversity. NatureServe, Arlington, Va.

Swannack, T. M., W. E. Grant, and M. R. Forstner, 2009. Projecting Population Trends of Endangered Amphibian Species in the Face of Uncertainty: A Pattern-oriented Approach. *Ecological Modelling* 220:148–159.

Texas Environmental Profile, 2008. *Wildlife Conditions in Texas.* http://www.texasep.org/html/wld/wld_3pna.html, accessed October 2008.

Texas Water Development Board, 2007. *Water for Texas.* Texas Water Development Board, Austin.

Tolan, J. M., 2007. El Niño–Southern Oscillation Impacts Translated to the Watershed Scale: Estuarine Salinity Patterns along the Texas Gulf Coast, 1982 to 2004. *Estuarine Coastal and Shelf Science* 72:247–260.

TPWD, 2003. *Texas Threatened and Endangered Species.* Texas Parks and Wildlife Department, Austin.

TPWD, 2005. *Texas Comprehensive Wildlife Conservation Strategy.* Texas Parks and Wildlife Department, Austin.

TPWD, 2010. *Land and Water Resources Conservation and Recreation Plan.* Texas Parks and Wildlife Department, Austin.

Twilley, R. R., E. J. Barron, H. L. Gholz, M. A. Harwell, R. L. Miller, D. J. Reed, J. B. Rose, E. H. Siemann, R. G. Wetzel, and R. J. Zimmerman, 2001. *Confronting Climate Change in the Gulf Coast Region: Prospects for Sustaining our Natural Heritage.* Union of Concerned Scientists and the Ecological Society of America, Cambridge, Mass.

U.S. Fish and Wildlife Service, 2008. *Bexar County Karst Invertebrates Draft Recovery Plan.* U.S. Fish and Wildlife Service, Albuquerque, N.M.

Vandergast, A. G., A. J. Bohonak, S. A. Hathaway, J. Boys, and R. N. Fisher, 2008. Are Hotspots of Evolutionary Potential Adequately Protected in Southern California. *Biological Conservation* 141:1648–1664.

Vitousek, P. M., 1989. Terrestrial Ecosystems. In: *Global Change and Our Common Future: Papers from a Forum*. R. S. DeFries and T. F. Malone (eds.). National Academy Press, Washington, D.C.

Wagner M., R. Kaiser, U. Kreuter, and N. Wilkins, 2007. Managing the Commons Texas Style: Wildlife Management and Ground-water Associations on Private Lands. *Journal of the American Water Resources Association* 43:698–711.

Wilcox, B. P., M. K. Owens, W. A. Dugas, D. N. Ueckert, and C. R. Hart, 2006. Shrubs, Streamflow, and the Paradox of Scale. *Hydrological Processes* 20:3245–3259.

Wilcox, B. P., L. P. Wilding, and C. M. Woodruff, 2007. Soil and Topographic Controls on Runoff Generation from Stepped Landforms in the Edwards Plateau of Central Texas. *Geophysical Research Letters* 34:L24S24. DOI: 10.1029/2007GL030860.

Williams, J. W., and S. T. Jackson, 2007. Novel Climates, No-analog Communities, and Ecological Surprises: Past and Future. *Frontiers in Ecology and the Environment* 5:475–482.

Williams, J. W., S. T. Jackson, and J. E. Kutzbach, 2007. Projected Distributions of Novel and Disappearing Climates by 2100 A.D. *Proceedings of the National Academy of Sciences* 104:5738–5742.

CHAPTER 6

Agriculture

Bruce A. McCarl

Productivity and income in agriculture are heavily influenced by climatic conditions. Changes in temperature, precipitation, extreme events, water flows, and atmospheric content have a mixture of positive and negative implications for plant growth, livestock performance, and water supply, as well as for soil characteristics, pests, and diseases. Thus, this industry is likely to face changing conditions and may be at risk given the possible incidence of global climate change. This chapter examines the vulnerability of agriculture in Texas to global climate change.

Estimation of the effects of climatic change on agriculture is difficult. Basically, there are three methods that could be used to make such an estimate. The first is based on observation, but it entails waiting for climate change conditions to develop fully, either globally or in representative regions. That is not now possible for the combination of carbon dioxide and climatic effects that we expect in the future. Second, one could turn to experimentation, by subjecting agricultural production systems to climatic change scenarios and observing the production implications. That is also not feasible, as the sites and systems to be investigated would render such an undertaking quite expensive. In addition, even if completed, the results would not reveal the effects on crop mix, markets, international trade, livestock herd size, and so on. That leaves the third, simulation-based approach, using models to simulate crop yields, crop mix choice, and market processes. This approach requires the adoption of scenarios regarding both climate effects and agricultural production and consumption conditions. It also necessitates the use of agricultural scenarios that are available, because the resources supporting this work do not permit reruns of the crop and hydrological simulations that are input. Thus the climate and associated agricultural and hydrological scenarios adopted are those resulting from the U.S. National Assessment (Reilly et al. 2001, 2002a,b).

The first part of the chapter discusses how global climate change might influence agricultural processes. Second, results are presented from an assessment of the influence of climate change on crop yields and the agricultural economy under 2007 conditions. The 2007 base was used, as opposed to a future year, because experience has shown that the model variation introduced by assumptions about future technological progress, demand, exports, and other factors is far larger than the implications of most phenomena such as climate change. Third, we focus on a regional study in the area around San Antonio, examining agricultural effects in the face of nonagricultural and ecological competition.

FACTORS DETERMINING THE SENSITIVITY OF AGRICULTURE

Agricultural production is influenced in numerous ways by the forces causing climate change, as well as the altered climate attributes. Drivers that lead to effects on agriculture will be grouped into five categories:

- Temperature affects plants, animals, pests, and water supplies. For example, temperature alterations directly affect crop growth rates, livestock performance and appetite, pest incidence, and water supplies in soil and reservoirs, among other influences.
- Precipitation alters the water directly available to crops, the drought stress crops are placed under, the supply of forage for animals, animal production conditions, irrigation water supplies, and river flows supporting barge transport, among other items.
- Changes in atmospheric carbon dioxide influence the growth of plants by altering the basic fuel for photosynthesis, as well as the water that plants need as they grow, along with the growth rates of weeds.
- Extreme events influence production conditions and water supplies, and they can alter waterborne transport and ports.
- A rise in sea level influences ports and waterborne transport, and it can inundate producing lands.

The reasons for such vulnerability and the manifestations of it are discussed in many references, including Adams et al. (1989, 1990, 1999), Reilly et al. (2002a), and the Intergovernmental Panel on Climate Change Reports (IPCC 2001, 2007).

NATIONAL AND TEXAS ANALYSIS

The climatic conditions projected for the future do not exist in today's world, and thus the implications of such changes cannot be fully evaluated. Some studies have examined differences in agricultural production patterns across regions with varying climates, using these differences to infer how shifts in climate would lead to shifts in production patterns. If, for example, the climate of Michigan became like that of Indiana, or North Dakota like that of South Dakota, then perhaps agriculture would shift in the northern states to be more like that of the more southern states. One of the major limitations of this procedure is that one cannot observe the consequences of an enriched carbon dioxide atmosphere, which will also have substantial effects on plant performance. Moreover, studies using such a procedure have not investigated effects of changing commodity prices and of international trade.

The approach here is to employ process models of crop and livestock response to weather and changing carbon dioxide concentrations and to simulate a market model of the agricultural economy. The process models can take advantage of the experimental evidence of crop response to changing carbon dioxide, and the market model can simulate price and trade effects of changing production. In doing this, we follow a number of previous studies and use a multistep analytical process:

- Projections about future atmospheric greenhouse gas emissions and concentrations are taken from existing studies.
- Climatologists use global circulation models to simulate future climatic conditions for the world, resolved at a latitude-longitude grid level, typically on the order of 2×2 degrees (areas of roughly 125 miles square).
- Agricultural and hydrological scientists employ crop, livestock, water flow, and other simulators to examine sensitivity as it varies geographically, scaling the climate projections from the circulation model down to sites where the detailed information needed to run the models is available. The simulators take into account the available water, soil characteristics, and other factors. They are typically simulated with management variables (such as fertilizer, water applied in areas with irrigation, and planting dates left) unchanged (no adaptation) and under changed conditions (with adaptation), either to increase yields further or to reduce yield losses.
- Economists use agriculture sector models to evaluate how multiple production changes, as they differ across the United States and the world, come together to affect markets, and how farmers in turn respond. The results of the economic modeling show how total crop and livestock production is

affected. Items of interest include shifts in the geographic incidence of crop acreage and livestock numbers, along with the effects on many performance measures, including markets, prices, incomes, and environmental factors. Such economic models include further adaptations that farmers would make in response to changing economic conditions. This would include changing the type of crop or livestock produced, abandoning or adding irrigation, or switching land into or out of crop production altogether.

This broad methodology will be employed herein. Only results from other studies will be used for the first three stages, regarding atmospheric greenhouse gas scenarios, climatic projections based on global circulation models, and simulations of water and agricultural crop and livestock performance. The new work done here involves the economic sector analysis. For water, crop, and livestock performance simulation data, we rely largely on the coordinated efforts of the agricultural component of the U.S. Global Climatic Change Research Program (USGCRP) of the U.S. national-level assessment (Reilly et al. 2001, 2002a,b). The Hadley and the Canadian climate scenarios are used here:

- The Hadley climate change scenario was developed by the United Kingdom Hadley Climate Center and used in the USGCRP National Assessment. According to the Agriculture Report (Reilly et al. 2002a), "For the continental United States, the Hadley scenario projects a 1.4°C (2030) and 3.3°C (2095) increase in temperature with precipitation increases of 6 and 23 percent, respectively . . . , more warming in the winter and relatively less in the summer. The mountain states and Great Plains are also projected to experience more warming than other regions . . . shows greater warming in the Northwest."
- The Canadian climate change scenario was developed by the Canadian Climate Center and was also used in the USGCRP National Assessment. According to the Agriculture Report (Reilly et al. 2002a), "For the continental United States, the Canadian model scenario projects a 2.1°C average temperature change with a 4 percent decline in precipitation by 2030 and a 5.8°C warming with a 17 percent increase in precipitation by 2095 . . . more warming in the winter and relatively

less in the summer. The mountain states and Great Plains are also projected to experience more warming than other regions."

Agricultural Sensitivity

A number of elements of agricultural production were shifted in the face of climate change. These and the procedures used are explained in McCarl and Reilly (2007).

- Crop yields were altered under climate change based on estimates from various crop simulators, as discussed in the U.S.
  agricultural sector appraisal (Reilly et al. 2002a). Data were
  generated for Texas regions for cotton, corn, and sorghum
  crops (Table 6.1). It is evident that the two climate change
  scenarios have implications for crop yields. Except for cotton,
  the Texas yield changes were generally lower or more negative than in the rest of the country.

TABLE 6.1. Percentage change in crop yields under two climate change scenarios for 2030, by Texas and U.S. region

	GLOBAL CLIMATE MODEL					
	HADLEY			CANADIAN		
TEXAS REGION	COTTON	CORN	SORGHUM	COTTON	CORN	SORGHUM
High Plains	101	−3	22	78	−14	43
Rolling Plains	46	9	24	26	0	53
Central Blacklands	36	11	24	21	2	54
East	12	9	23	9	1	51
Edwards Plateau	35	−2	23	21	−12	48
Coastal Bend	11	11	24	6	3	55
South	20	6	23	13	−3	47
Trans-Pecos	9	−4	17	7	−15	13
U.S. REGION						
Corn Belt	13	12	36	8	12	34
Great Plains	13	9	30	8	6	25
Lake States	0	47	0	0	62	0
Northeast	0	5	33	0	0	28
Rocky Mountains	31	4	87	26	−2	52
Pacific Southwest	−6	−3	107	−5	−7	53
Pacific Northwest	0	−3	0	0	−7	0
South Central	20	7	30	13	−5	26
Southeast	21	4	29	12	−8	25
Southwest	69	4	26	50	−5	43
U.S. average	34	16	31	24	16	33

- Levels of inputs such as fertilizer, energy, labor, and insecticides were varied with crop production changes. For example, if yields are higher, greater inputs are needed, whereas if yields are lower, then less are needed. Farm-level evidence suggests that the change in input use is less than proportional to the yield change. The estimated relationships vary by crop, but for most crops the change in input use was on the order of 0.4 percent for a 1.0 percent change in yield.
- Nationally, water demand by irrigated crops dropped substantially for most crops, but it increased in Texas.
- Climate change can have implications for livestock, principally through changes in appetite and the distribution of energy between maintenance and growth. Disease incidence is also likely to be affected. The result is altered milk and meat production, meat quality, and species reproduction as climate changes. A study sponsored by the Electric Power Research Institute (Adams et al. 1999) developed relationships between temperature change, livestock performance, and feedstuff consumption in consultation with experts on livestock production and management. The results generally show a reduction of 5–7 percent in annual per animal production of meat, milk, and wool.
- The amount of feedstuffs and other inputs changes when livestock productivity changes. We assume that feedstuff use is strictly proportional to the volume of animal products produced. The use of nonfeed inputs changed by 0.5 percent for every 1.0 percent change in livestock yields.
- Climate change can affect water supply and, in turn, the amount of irrigation water available for agriculture in several ways. The national assessment water study (Gleick 2000) developed a set of climate effects on surface water availability. These data show that the Pacific Southwest gains the most (this might be modified under the latest climate scenarios, since that region is drier than in the circulation model runs used here) under the climate change scenarios, with the smallest gains and largest losses generally being in the southern regions, including Texas. Note that the Canadian scenario is the most extreme, and the Hadley is the most optimistic.

- Climate change will affect grass growth and thus the effective supply of pasture and animals that can be supported on western grazing lands. We assumed that climate change altered livestock usage of grazing lands in proportion to the effect of climate change on animal performance. We also assumed altered rates of grass growth, as the changing climate effectively changed grazing land availability using the hay crop simulation results.

- The evidence suggests that problems involving pests (herein defined as insects, weeds, and diseases) are greater in warmer areas. Thus, climate change may lead to changes in the range and incidence of agriculturally damaging pests. To consider how climate could affect agriculture, we measured pest damage as a change in expenditures on pest control, an approach that was developed in the U.S. National Assessment (Chen et al. 2001). The results show an increase in crop expenditures for almost all crops under all scenarios. The largest increases are found for corn and potatoes, with smaller increases for cotton and the wheat crops.

Economic Analysis

The agricultural sector scenario assumes the 2007 status quo, as projection of future rates of technological progress and consumption growth is both difficult and potentially a much larger factor than climate change itself, thereby swamping the climate change effects. The year 2007 is used throughout.

Three economic model runs were performed, including a base case without climate change and a case for each climate change scenario. The overall societal welfare results show total increases of about $29 billion under the Hadley scenario and about $23 billion under the Canadian scenario (Table 6.2). In addition, producers benefit under both scenarios, obtaining most of the gain, whereas consumer welfare shows smaller but positive gains. Foreign welfare shows gains, but with losses to foreign producers. These results suggest that the national agricultural sensitivity to global climate change is small. The growth-stimulating effects of carbon dioxide, coupled with cropping pattern substitution, overcome the negative effects regarding water supply and, in some cases, yield.

When model projections regarding commodity prices and production are reviewed, price decreases are observed across almost all commodities, and these price decreases are matched by increased production levels. In Texas

TABLE 6.2. National and global societal welfare results under two climate change scenarios

AGRICULTURAL SECTOR	BASE	CLIMATE SCENARIO	
		CANADIAN	HADLEY
		CHANGE FROM THE BASE, IN MILLIONS OF DOLLARS	
United States			
Consumers	1,781,612	1,772	9,799
Processors	3,249	282	121
Producers	66,911	18,143	16,548
Total	1,851,772	20,195	26,468
Rest of the World			
Consumers	289,973	3,782	4,741
Producers	20,806	−991	−1,805
Total	310,780	2,792	2,936
Total Globally	2,162,552	22,988	29,404

(Table 6.3) more cotton and sorghum are projected, while production of wheat, corn, rice, cattle, and broilers falls. Cropped acres are reduced by about 20 percent, and use of irrigation water falls largely because less water is available.

EDWARDS AQUIFER REGIONAL ANALYSIS

A very important issue regarding climate change and Texas agriculture involves the trade-offs between agricultural, municipal, industrial, and environmental uses of water. The results of a study by Chen et al. (2001) regarding the Edwards Aquifer of San Antonio, Texas, provide a review of such

TABLE 6.3. Change in selected Texas commodities under two climate change scenarios

COMMODITY	BASE	SCENARIO	
		CANADIAN	HADLEY
Producer net income (millions of dollars)	4,757	4,707	4,253
Production index			
All farm products	100.00	90.78	96.54
All crops	100.00	90.70	96.46
All livestock	100.00	90.15	96.00
Price index			
All farm products	100.00	101.61	93.12
All crops	100.00	92.85	91.10
All livestock	100.00	107.13	94.56
Calves in feedlots (1,000 head)	8,095	7,000	771
Total broilers (1,000 animals)	596,066	488,384	542,917
Acres cropped	1,267,426	985,952	984,254
Irrigation water use (1,000 acre-feet)	5,831	5,885	5,500

issues. In particular, that paper explored the implications of recent climate change projections for the aquifer region, concentrating on changes in water use patterns among the sectors, environmental matters, and the economy.

There are many competing uses for the waters of this aquifer, including municipal, agricultural, industrial, military, and recreational. The Edwards also discharges water to artesian springs in the eastern part of the aquifer. Pumping in the western part is largely by agricultural users, whereas eastern pumping is mainly by municipal and industrial water users. Spring discharge, mainly from the San Marcos and Comal springs in the east, supports a habitat for endangered species (Longley 1992), provides water for recreational use, and serves as an important supply source for water users in the Guadalupe-Blanco river system. The Edwards Aquifer has substantial recharge, averaging 658,200 acre-feet between 1934 and 1996. Over that period, pumping and springflow discharge averaged 668,700 acre-feet (U.S. Geological Survey 1997). Use has increased in the past, with pumping rising by 1 percent a year in the 1970s and 1980s (Collinge et al. 1993) and, as of 2001, accounted for 70 percent of the total discharge. The increasing use placed the aquifer under stress, leading to increased annual variability, lessened springflow, and increased concern over the endangered species habitat. This stress and other factors led to a successful lawsuit by the Sierra Club to protect the endangered species (Bunton 1996) and to action by the Texas Legislature placing the aquifer under pumping limitations (Texas Senate 1993), although these were increased in 2007.

Climate change could increase the stress on the aquifer, altering recharge and increasing water demand. This study utilizes an existing hydrological and economic systems model for the Edwards, the EDSIM (McCarl et al. 1998), coupled with an examination of climate change implications for recharge and water demand.

Climatic Change Recharge and Water Demand

The results from U.S. National Assessment scenarios were used to provide the climate change scenarios. The average changes in regional temperature and precipitation for 10-year periods are centered on 2030 and 2090 (Table 6.4). Such changes would alter water demand and supply. The implications for aquifer recharge, crop and municipal water demand, and agricultural yields are examined here.

Recharge Implications

Chen et al. used a regression analysis to estimate the effects of climate change, in the form of changes in temperature and precipitation, on recharge.

TABLE 6.4. Changes in Edwards Aquifer region temperature and precipitation, by climate change scenario

CLIMATE CHANGE SCENARIO	TEMPERATURE CHANGE (°F)	PRECIPITATION CHANGE (INCHES)
Hadley 2030	3.20	−4.10
Hadley 2090	9.01	−0.78
Canadian 2030	5.41	−14.36
Canadian 2090	14.61	−4.56

In particular, U.S. Geological Survey (1997) estimates of historical recharge data by county from 1950 to 1996 were drawn from annual reports of the Edwards Aquifer Authority; associated climate data were obtained from the Office of the Texas State Climatologist (1998) and a University of Utah Web page.

Monthly recharge was forecast as a log-linear function of temperature and precipitation, as explained by Chen et al. The summary measures for recharge implications show recharge reductions ranging from 19.68 to 48.86 percent (Table 6.5).

Municipal Water Use Implications

To estimate how municipal water demand would be affected, Chen et al. relied on estimates from Griffin and Chang (1991). They adjusted the daily climate record from 1950 to 1996, altering the temperature and precipitation so that it reflected the differences in the climate change scenarios. The results show that climate change would increase municipal water demand by 1.5–3.5 percent.

TABLE 6.5. Selected effects of scenarios on Edwards Aquifer region

EFFECT	CHANGE FROM BASE (PERCENT)			
	HADLEY 2030	HADLEY 2090	CANADIAN 2030	CANADIAN 2090
Aquifer recharge				
Recharge in drought years	−20.59	−32.89	−29.65	−31.96
Recharge in normal years	−19.68	−33.46	−28.99	−36.23
Recharge in wet years	−23.64	−41.45	−34.42	−48.86
Municipal water demand	1.539	2.521	1.914	3.468
Crop yield and water use				
Irrigated corn yield	−1.93	−3.47	−4.26	−5.61
Irrigated corn water use	11.95	31.32	23.47	54.03
Dryland corn yield	−3.93	−6.78	−8.17	−10.79
Irrigated sorghum yield	−1.75	−3.35	−2.79	−4.17
Irrigated sorghum water use	15.12	38.16	42.65	79.36
Dryland sorghum yield	−5.93	−13.07	−10.82	−16.76
Irrigated cotton yield	−9.06	−15.82	−19.80	−24.64
Irrigated cotton water use	16.88	40.82	34.58	71.50

Crop Yields and Irrigation Water Use

Climatic change would also influence crop yields and irrigation requirements. Chen et al. estimated this using the Blaney-Criddle procedure (Heimes and Luckey 1983; Doorenbos and Pruitt 1977). Summary measures of the resultant effects show a decrease in yields and an increase in water requirements (Table 6.5).

Economic Damages

The climate-induced changes estimated above were factored into the regional Edwards Aquifer economic and hydrological simulation model, EDSIM, developed by Dillon (1991), McCarl et al. (1993), Lacewell and McCarl (1995), Keplinger and McCarl (1995), Keplinger (1996), Williams (1996), and McCarl et al. (1998). EDSIM depicts pumping use by the agricultural, municipal, and industrial sectors and simultaneously calculates pumping lift, ending elevation, and springflow. EDSIM simulates regional municipal, industrial, and agricultural water use, irrigated versus dryland production, and choice of irrigation delivery system (sprinkler or furrow) across a nine-state representation of the probability distribution of precipitation, aquifer recharge, and crop water demand and yield. The model computes regional welfare, which is the sum of net farm income and municipal and industrial consumer surplus. Total water usage is constrained to be less than or equal to the pumping limit of 400,000 acre-feet, as mandated by the Texas Senate for years after 2008.

The five scenarios considered by Chen et al. (2001) were (1) the base, without climate change, (2) climate change as predicted by the Hadley model for 2030, (3) climate change as in the Canadian model for 2030, (4) climate change as in Hadley for 2090, and (5) climate change as in Canadian for 2090.

Among the results, the strongest effect of climate change falls on springflow and the agricultural sector (Table 6.6). Springflows at Comal (the most sensitive spring) decrease 10–17 percent under the 2030 scenarios and 20–24 percent under 2090. Farm income falls 16–30 percent under the 2030 scenarios and 30–45 percent under 2090. The shift in agricultural water use to municipal and industrial uses indicates that the city users will buy out some agricultural usage through water markets.

Despite an increase in municipal and industrial water use, the municipal-industrial surplus decreases, as lower recharge and falling aquifer levels cause an increase in pumping lift and cost. The value of water use permits increases by 5–24 percent. Water use in the nonagricultural sector is less

TABLE 6.6. Edwards Aquifer regional results under alternative climate change scenarios

			PERCENT CHANGE FROM BASE SCENARIO			
VARIABLE	UNITS	BASE	HADLEY 2030	HADLEY 2090	CANADIAN 2030	CANADIAN 2090
Agricultural water use	thousands of acre-feet	150.05	−0.89	−2.4	−1.35	−4.15
Municipal-industrial water use	thousands of acre-feet	249.72	0.63	1.54	0.9	2.59
Total water use	thousands of acre-feet	399.77	0.06	0.06	0.06	0.06
Net farm income	thousands of dollars	11,391	−15.85	−30.34	−29.41	−44.97
Net municipal-industrial surplus	thousands of dollars	337,657	−0.2	−0.58	−0.36	−0.92
Authority surplus[a]	thousands of dollars	6644	3.76	12.73	7.07	21.6
Net total welfare	thousands of dollars	355,692	−0.64	−1.3	−1.16	−1.93
Total annual flow at Comal Springs	thousands of acre-feet	379.5	−9.95	−20.15	−16.62	−24.15
Total annual flow at San Marcos Springs	thousands of acre-feet	92.8	−5.07	−10.09	−8.3	−12.06

[a] Welfare accruing to the pumping or springflow limit; the rental value of all permits
Source: Chen et al. (2001)

variable; a shift to that sector actually makes water use slightly greater, with corresponding declines in springflow.

The reduction in springflow would put the endangered species habitat in additional peril. Thus, a reduction in the allowed pumping may be required to protect the springs, endangered species, and other environmental amenities. Additional simulations show that the level of the aquifer pumping limit needs to decrease by 9 percent, to maintain the springflow level, at an additional cost of $0.5–$2 million per year.

CONCLUSIONS

A quantitative examination of the agricultural effects of climate change was carried out in a nationwide context. Generally, the results show (much as they did in the first edition) that agriculture in the United States and Texas is sensitive in terms of land and water usages, as well as crop and livestock production. In terms of agriculture-based economic welfare, however, the simulated effects of climate change are not large. There will be regional displacements; Texas agriculture is vulnerable to climate change as simulated here, particularly in the High Plains. Under these climate change conditions, the simulation also shows that statewide cropped acreage declines by about 20 percent.

The nature of these results, particularly the overall resilience of agriculture to climate change, is not entirely unexpected. The pattern is similar to the results of many earlier studies. The Edwards Aquifer study also shows that agriculture, water competition among sectors, and ecology are at issue.

REFERENCES

Adams, R. M., J. D. Glyer, and B. A. McCarl, 1989. The Economic Effects of Climate Change on U.S. Agriculture: A Preliminary Assessment. In: *The Potential Effects of Global Climate Change on the United States*. Report to Congress by the U.S. Environmental Protection Agency, Washington, D.C.

Adams, R. M., C. Rosenzweig, R. M. Peart, J. T. Ritchie, B. A. McCarl, J. D. Glyer, R. B. Carry, J. W. Jones, K. J. Boote, and L. H. Allen. 1990. Global Climate Change and U.S. Agriculture. *Nature* 345:219–224.

Adams, R. M., B. A. McCarl, K. Segerson, C. Rosenzweig, K. J. Bryant, B. L. Dixon, J. R. Conner, R. E. Evenson, and D. S. Ojima, 1999. The Economic Effects of Climate Change on U.S. Agriculture. Pp. 18–54 in: *The Impact of Climate Change on U.S. Agriculture*. R. Mendelsohn and J. Newmann (eds.). Cambridge University Press, London.

Bunton, Lucius D. III, 1996. *Order on the Sierra Club's Motion for Preliminary Injunction, Sierra Club et al. v. Bruce Babbitt et al.* MO-91-CA-069, United States District Court, Western District of Texas, Midland-Odessa Division.

Chen, C. C., D. Gillig, and B. A. McCarl, 2001. Effects of Climatic Change on a Water Dependent Regional Economy: A Study of the Texas Edwards Aquifer. *Climatic Change* 49:397–409.

Collinge, R., P. Emerson, R. C. Griffin, B. A. McCarl, and J. Merrifield, 1993. *The Edwards Aquifer: An Economic Perspective*. TR-159. Texas Water Resources Institute, Texas A&M University, College Station.

Dillon, C. R., 1991. An Economic Analysis of Edwards Aquifer Water Management. Dissertation, Texas A&M University, College Station.

Doorenbos, J., and W. O. Pruitt, 1977. *Guidelines for Predicting Crop Water Requirements*. FAO Irrigation and Drainage Paper 33. Food and Agriculture Organization of the United Nations, Rome.

Gleick, P. H. (lead author), 2000. *Water: The Potential Consequences of Climate Variability and Change for the Water Resources of the United States*. Report of the National Water Assessment Group, U.S. Global Change Research Program. Pacific Institute for Studies in Development, Environment, and Security, Oakland, Calif.

Griffin, R. C., and C. Chang, 1991. Seasonality in Community Water Demand. *Western Journal of Agricultural Economics* 16:207–217.

Heimes, F. J., and R. R. Luckey, 1983. Estimating 1980 Groundwater Pumpage for Irrigation on the High Plains in Parts of Colorado, Kansas, Nebraska, New

Mexico, Oklahoma, South Dakota, Texas, and Wyoming. Water Resources Investigation Report 83–4123. U.S. Geological Survey, Denver, Colo.

IPCC, 2001. *Climate Change 2001: Impacts, Adaptation, and Vulnerability.* Contribution of Working Group 2 to the Third Assessment Report of the Intergovernmental Panel on Climate Change. Cambridge University Press, Cambridge, U.K., and New York. http://www.ipcc.ch/ipccreports/tar/wg2/index.htm.

IPCC, 2007. *Climate Change 2007: Impacts, Adaptation, and Vulnerability.* Contribution of Working Group 2 to the Fourth Assessment Report of the Intergovernmental Panel on Climate Change. Cambridge University Press, Cambridge, U.K., and New York. http://www.ipcc.ch/ipccreports/ar4-wg2.htm.

Keplinger, K. O., 1996. An Investigation of Dry Year Options for the Edwards Aquifer. Dissertation, Texas A&M University, College Station.

Keplinger, K. O., and B. A. McCarl, 1995. Regression Based Investigation of Pumping Limits on Springflow within the Edwards Aquifer. Manuscript. Department of Agricultural Economics, Texas A&M University, College Station.

Lacewell, R. D., and B. A. McCarl, 1995. *Estimated Effect of USDA Commodity Programs on Annual Pumpage from the Edwards Aquifer.* Final report submitted to the U.S. Department of Agriculture, Natural Resource Conservation Service, Temple, Tex.

Longley, G., 1992. The Subterranean Aquatic Ecosystem of the Balcones Fault Zone Edwards Aquifer in Texas: Threats from Overpumping. Pp. 291–300 in: *Proceedings of the First International Conference on Ground Water Ecology.* J. A. Stanford and J. J. Simons (eds.). Ecological Society of America, Tampa, Fla.

McCarl, B. A., and J. M. Reilly, 2007. U.S. Agriculture in the Climate Change Squeeze, Part 1: Sectoral Sensitivity and Vulnerability. Report to National Environmental Trust. http://agecon2.tamu.edu/people/faculty/mccarl-bruce/papers/1303Agriculture in the climate change squeez1.doc.

McCarl, B. A., W. R. Jordan, R. L. Williams, L. L. Jones, and C. R. Dillon, 1993. *Economic and Hydrologic Implications of Proposed Edwards Aquifer Management Plans.* Technical Report 158. Texas Water Resources Institute, Texas A&M University System, College Station.

McCarl, B. A., K. O. Keplinger, C. Dillon, and R. L. Williams, 1998. Limiting Pumping from the Edwards Aquifer: An Economic Investigation of Proposals, Water Markets, and Springflow Guarantees. *Water Resources Research* 35(4):1257–1268.

Office of the Texas State Climatologist, 1998. Web page http://www.met.tamu.edu/met/osc/osc.html, accessed April 1998.

Reilly, J. M., F. Tubiello, B. A. McCarl, and J. Melillo, 2001. Climate Change and Agriculture in the United States. In: *Climate Change Impacts on the United States: U.S. National Assessment of the Potential Consequences of Climate Variability and Change.* Cambridge University Press, Cambridge, U.K. http://www.gcrio.org/nationalassessment/, 379–403.

Reilly, J. M., J. Hrubovcak, J. Graham, D. G. Abler, R. Darwin, S. E. Hollinger, R. C. Izaurralde, S. Jagtap, J. W. Jones, J. Kimble, B. A. McCarl, L. O. Mearns,

D. S. Ojima, E. A. Paul, K. Paustian, S. J. Riha, N. J. Rosenberg, C. Rosenzweig, and F. Tubiello, 2002a. *Changing Climate and Changing Agriculture: Report of the Agricultural Sector Assessment Team, U.S. National Assessment.* Prepared as part of the USGCRP National Assessment of Climate Variability, Cambridge University Press, Cambridge, U.K.

Reilly, J. M., F. Tubiello, B. A. McCarl, D. G. Abler, R. Darwin, K. Fuglie, S. E. Hollinger, R. C. Izaurralde, S. Jagtap, J. W. Jones, L. O. Mearns, D. S. Ojima, E. A. Paul, K. Paustian, S. J. Riha, N. J. Rosenberg, and C. Rosenzweig, 2002b. U.S. Agriculture and Climate Change: New Results. *Climatic Change* 57:43–69.

Texas Senate, 1993. Senate Bill 1477, 73rd Session. Austin.

U.S. Geological Survey, 1997. *Recharge to and Discharge from the Edwards Aquifer in the San Antonio Area, Texas.* U.S. Geological Survey, Austin, Tex. http://tx.usgs.gov/reports/district/98/01/index.html.

University of Utah, 1998. Web page, http://climate.usu.edu/ accessed April 1998.

Williams, R. L., 1996. Drought Management and the Edwards Aquifer: An Economic Inquiry. Dissertation, Texas A&M University, College Station.

CHAPTER 7

Urban Areas
David Hitchcock

Texas cities and the people who live there are vulnerable to the effects of global climate change because of the heavy concentration of population, infrastructure, and economic activities. The concentration of human, community, and capital resources also means that cities are where the potential to adapt is the strongest, and where system changes (such as infrastructure redevelopment) require more vision. This chapter describes the vulnerabilities of major Texas cities to climate change impacts and the variations in those impacts from city to city due to their location within the state and subregional climate conditions. *Vulnerable* is used here to indicate characteristics of locations that could result in climate change impacts; for example, Gulf Coast cities are vulnerable to sea-level rise.

According to the most recent reports of the Intergovernmental Panel on Climate Change (IPCC), cities are now expected to experience localized impacts from higher average and peak temperatures, higher nighttime temperatures, changes in stormwater runoff, increased precipitation (5–25 percent), and more frequent and intense storms linked to urban expansion (Wilbanks et al. 2007). The projections, however, are based on climate models that do not simulate the global and regional effects due to urbanization, such as land use and land cover alterations (Christensen et al. 2007). Therefore, climate change effects of primary importance to cities cannot be projected with the same tools as climate phenomena. Responses to climate change vulnerabilities rely on technological and institutional changes that occur over several decades. These changes are the products of human and societal invention that result from discoveries made in the context of uncertain future conditions. Climate assessments, as they affect cities, must view the future in terms of vulnerabilities rather than projections (Wilbanks et al. 2007:364).

Communities and community leaders are faced with the worst kind of dilemma, that of taking action on the basis of uncertainties that have potentially severe consequences. As such, it has been common in Texas to deny that there is a problem or to choose inconsequential actions. In one reported exchange, regional leaders suggested that the phrases "climate change" or "global warming" should not be used but instead replaced with "environmental effects" to "avoid the whole controversy" (*Houston Chronicle*, November 27, 2007). As of this writing, only a few Texas cities and community leaders have acknowledged that climate change poses a threat and have become willing to discuss issues and possible responses.

Cities have strong interdependencies between the physical infrastructure (transportation, energy, buildings, and communications) and societal infrastructure (services and economic, political, governmental, institutional, and health systems). Texas was transformed during the twentieth century from a largely rural setting to an area in which three quarters of its citizens and most of its economy are urban. Today's Texans live and work in cities, and the future portends an even larger urban setting. Over time, these cities have adapted to Texas climate conditions, responding to the historical challenges of rainfall, drought, floods, heat waves, high temperatures, ice storms, tornadoes, and hurricanes. Cities rely on physical and societal infrastructures that are vulnerable to global climate change impacts (Ruth and Kirshen 2002). The wide-ranging climate conditions across Texas also mean that these vulnerabilities occur differently from place to place.

The largest Texas urban areas are home to more than 18 million people. The largest urban concentrations include Dallas–Fort Worth, Houston, San Antonio, Austin, El Paso, the urbanized areas of the Lower Rio Grande Valley, and communities along the Texas Gulf Coast (Fig. 7.1). Those cities with a hot, arid climate, such as El Paso, already face water resource challenges and are vulnerable in the future to a hotter, drier climate that will exacerbate these conditions. Gulf Coast areas, including Houston, face the prospect of more frequent and severe storms, as well as the threat of a rising sea level. Central and North Texas cities (including Austin, San Antonio, and the Dallas–Fort Worth area) are not directly vulnerable to sea-level rise, but they will experience the consequences of more frequent and severe storms generating from the Gulf Coast. Texas cities are projected to continue their pattern of growth and expansion, making them increasingly vulnerable to water-supply shortages and the impacts of increased energy demand.

Major cities have greater capacity than small communities and rural areas to adapt to the consequences of climate change because of their relative wealth, infrastructure, financing capability, health systems, and governance capabilities. Governance in Texas urban regions involves hundreds of local governments of various sizes and capabilities. There are also hundreds of special districts that manage and control various functions, such as wastewater treatment. Statewide decision authority on many urban climate issues

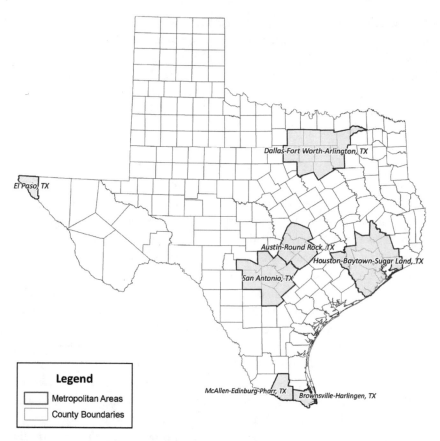

FIGURE 7.1. Metropolitan statistical areas of Texas.

largely resides at the state government level, but in Texas this system must respond to a large and highly diverse set of political interests that seek to limit the government's role. While providing broader geographic coverage than municipalities, county governments in Texas have very limited authority. In urban Texas, regional authorities are limited primarily to planning activities and cooperative efforts involving several jurisdictions. They lack the regulatory authority to respond to issues such as those posed by climate change. To reach major decisions, large Texas cities have historically relied on business and community leaders rather than elected officials and local government. This has changed somewhat in recent years.

Cities are also constrained by the inertia of their investments in urban residential patterns, water and sewer infrastructure, and transportation, as well as electric power systems. Such systems are difficult and expensive to retrofit or adapt. Additionally, as demonstrated in disasters such as

Hurricane Katrina, and under disparate economic or environmental conditions, people will migrate to cities seeking better conditions. Major Texas cities must be prepared for such migrations.

FACTORS CONTRIBUTING TO URBAN VULNERABILITY TO CLIMATE CHANGE
Population Growth

More than two thirds of the Texas population lives and works in the six major urban centers of Austin–Round Rock, Dallas–Fort Worth–Arlington, El Paso, Houston–Sugar Land–Baytown, the Lower Rio Grande area (including Brownsville and McAllen), and San Antonio. By 2040, these areas are projected to become a larger fraction (71 percent) of the state's total population. More revealing, these six areas will account for all but 16 percent of the state's population growth (Table 7.1).

By 2040, the Dallas–Fort Worth and Houston areas would add 7.7 million people (4.4 million and 3.3 million, respectively). Other urban centers along the Gulf Coast, including Beaumont–Port Arthur and Corpus Christi, are expected to grow as well. The Lower Rio Grande area is projected to have the largest percentage increase, at 105 percent, followed by Austin at 89 percent. Over the last century, the widespread rural population of Texas in the 1900s has moved from rural counties to the current major urban centers (Fig. 7.2).

As suggested here, population growth in Texas cities will be a major determinant of the need for responses to future climate change. Cities can respond proactively to this growth in ways that could accommodate expected climate change effects, including provisions within infrastructure standards, building standards, and development controls. As providers of

TABLE 7.1. Population of Texas and major SMSAs, 2005 and 2040

AREA	2005	2040
Texas	22,556,054	35,761,201[a]
Major SMSAs		
Austin–Round Rock	1,405,638	2,658,510
Dallas–Fort Worth–Arlington	5,667,966	10,106,814
El Paso	740,648	1,153,058
Houston–Sugar Land–Baytown	5,121,006	8,400,148
Lower Rio Grande	1,032,392	2,115,220
San Antonio	1,833,252	2,512,021
Metro area total (SMSAs)	15,800,902	26,945,771
Rest of state total	6,755,152	8,815,430

[a] Water Development Board population projections for 2040 are 36,893,267, 3 percent more than Texas State Data Center projections for 2040.

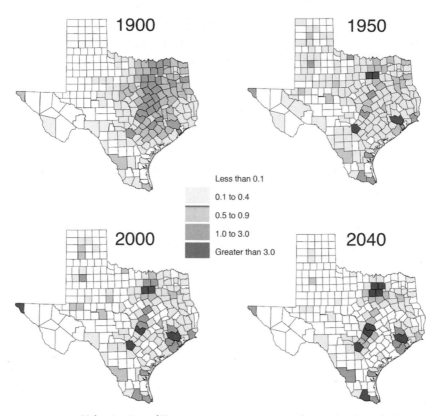

FIGURE 7.2. Urbanization of Texas 1900, 1950, 2000, and 2040 (projected). Percentage of state population by county. *Source:* Texas State Data Center

public health and public safety, cities will also need to respond in these areas. Texas urban centers face a range of vulnerabilities, but they will also discover some common elements that can be put in place regardless of the particular vulnerability. For example, energy efficiency measures, whether applied to buildings or vehicles, work across the board to reduce impacts and improve response capabilities. Longer public sector investment timelines for infrastructure (i.e., 50-year life rather than 20-year life) could improve infrastructure resilience while reducing overall life-cycle costs. Such public investment decisions can also better incorporate environmental and other externalities relevant to climate change.

Climate Diversity

The major urban centers of Texas are found across a wide range of climatic conditions, from the hot, arid climate in Far West Texas to the humid, subtropical conditions in Houston. These variations suggest that adaptation strategies will differ from city to city. Average annual precipitation varies across these cities by almost 40 inches. Of the six areas, only Austin, San Antonio, and Houston have the same building energy code, which is based on their climate region. The other three urban areas are each in different climate regions (Table 7.2).

TABLE 7.2. Climate features of major Texas urban areas

URBAN AREA	CLIMATE REGION[a]	AVERAGE RAINFALL INCHES/YEAR	AUGUST AVERAGE HIGH-LOW TEMPS °F[b]	JANUARY AVERAGE HIGH-LOW TEMPS °F[b]	ENERGY CODE REGION[c]
Austin–Round Rock	South Central	33.6	96-73	60-40	4
Dallas–Fort Worth–Arlington	North Central	34.7	95-74	54-34	5
El Paso	Far West	9.4	92-70	57-33	6
Houston–Sugar Land–Baytown	Southeast	47.8	94-73	62-41	4
Lower Rio Grande	South	27.6	93-75	69-50	2
San Antonio	South Central	32.9	95-74	62-39	4
Range or variation among areas		9.4–47.8	4°F–5°F	15°F–17°F	

[a] As in Figure 2.4, in Chapter 2
[b] National Weather Service data
[c] Texas Residential Building Guide to Energy Code Compliance, May 1, 2001. Energy Systems Laboratory, Texas A&M University

Severe weather (such as hail, wind, and tornadoes) occurs more frequently in the Dallas–Fort Worth area than in the five other urban areas, with lower frequencies in Brownsville and El Paso. Houston and the Lower Rio Grande are more vulnerable to hurricanes and tropical storms.

Although climate variables such as maximum hourly rainfall are accounted for in urban development and building codes, the climate variations and new extremes projected with global climate change are not. Cities are built based on historical climate conditions and, to a certain degree, historical climate extremes. If climate changes are not somehow anticipated in such codes and urban policies, adaptation measures are slowed by the delay in initial responses. Planners and government agency staff may be concerned about climate change in Texas, but without decisions by city leaders, they are not openly incorporating such factors in relevant activities. As such, climate-relevant scenarios are omitted from planning for urban functions such as stormwater infrastructure, transportation facilities, air-quality planning, building codes, and water resources.

Clearly, one-size-fits-all responses will not work within the climate variations across Texas cities. Cities do need a common framework, including data and forward-looking projections, to incorporate climate change in their planning.

Urban Heat Islands

Because of the urban heat island effect, cities are hotter than rural surroundings by several degrees (up to 10°F). With temperature increases induced by climate change, this effect will be intensified. Urban heat island effects occur when vegetative cover is reduced or removed for urban development and replaced with heat-absorbing materials, such as dark paving and roof surfaces. The results include not only higher temperatures but also changes in other climate variables such as soil moisture, rainfall patterns, wind direction and intensity, and lightning strikes.

Urban heat island effects can be divided into three layers: the canopy, the boundary, and the surface. The urban canopy extends from the ground surface to the height of the average building. The boundary layer extends up to 1 kilometer, expanding with heat during the day and contracting during the cooler nighttime. The urban heat effect is often pictured as a dome of air, or island, over the city. The surface effects are associated with urban surface characteristics such as hot roofing surfaces, which can reach 180°F during hot summer days.

Paving and roofing account for over 60 percent of the surface of developed urban areas. Higher urban temperatures are associated with higher average and peak energy demand for cooling and with higher levels of ozone. In Houston, half of the urban surface is paving (29 percent) and roofing (21 percent) (Rose et al. 2003). As cities develop, the extent of the urban heat island increases (Oke 1973; Streutker 2002). Vegetative cover can reduce urban temperatures through transpiration (evaporative cooling from released moisture) and shade. Trees and buildings also affect wind flow and wind patterns in cities. Human-induced heating in cities also comes from motor vehicle engines, air-conditioning condensers, building cooling towers, generators, power plants, and industrial processes.

Studies of urban temperature change show that average daytime temperatures have increased in the past by 0.2–0.8°F per decade (Akbari 2005). At Dallas Love Field, average wintertime temperatures increased by 0.8°F per decade from 1970 to 2006 while summertime temperatures increased by 0.4°F per decade. These increases occurred as the surrounding area became urbanized. Houston Hobby Airport experienced increases as well, but at a much lower rate (less than 0.1°F per decade).

Over the past 50 years, cities in the southern tier of the United States experienced intensification of the urban heat island effect as measured by differences in urban and rural temperatures (Stone 2007). Average daily temperature increases over 0.9°F per decade were observed in Oklahoma City, and heat-island intensity increased in Dallas, Austin, and San Antonio, with Dallas and Austin experiencing increases of 0.4–0.5°F per decade. The intensity of the heat island was lessened in the Houston area, perhaps because of the moderating effects of the Gulf Coast climate. Urban heat island effects also occur differently from city to city. For example, in the Dallas–Fort Worth area, the nighttime effect of the heat island is more prominent than it is in the Houston area (Darby and Senff 2007).

The effects and mitigation of urban heat islands are relatively well understood, although not extensively applied. The basic ways of reducing the effects are increased vegetative cover (more trees), reduced heat-absorbing and impervious surfaces (less paving, use of pervious paved surfaces), and increased solar reflectance of urban surfaces (lighter paving and roofing). California was the first state to pursue cool roof technologies by including them in the statewide building code. Despite the availability of methods and technologies to reduce urban temperatures, most heat-island initiatives have been incremental, rather than systemic. Cities to date have been unwilling to take on the comprehensive measures required to change urban surfaces sufficiently to minimize urban heat island effects (Hitchcock 2004; Environmental Protection Agency 2008).

CHALLENGES FACING POLICYMAKERS
Water Demand
Texas urban centers are a dominant factor in future water demand. The shift from rural to urban water uses will continue far into the future (Texas Water Development Board 2007). Urban uses such as electric power generation (steam electric) and manufacturing growth are also projected to drive future water demand. Decreased precipitation, reduced stream flows, and increased evaporation from reservoirs due to climate change will hamper the state's ability to meet future urban water needs, particularly in those areas where population expansion and economic growth will occur—the major Texas cities.

In the past, Texas water management policies focused primarily on reservoir development and greater reliance on surface waters. The state's regional planning process has changed to include methods such as conservation, reuse, and desalination (although many of these measures are pushed far into the future). This is particularly evident in water regions containing major urban centers. Since population growth is the largest single determinant, future demand will come largely from urban regions (Table 7.3).

TABLE 7.3. Population and water demand in urban and nonurban regions

REGION[a]	2010		2040		% CHANGE	
	POPULATION, MILLIONS	WATER DEMAND, MILLIONS OF ACRE-FEET	POPULATION, MILLIONS	WATER DEMAND, MILLIONS OF ACRE-FEET	POPULATION, MILLIONS	WATER DEMAND, MILLIONS OF ACRE-FEET
Urban regions	18.66	7.37	28.87	10.15	54.7	37.8
Nonurban regions	5.77	5.08	7.48	5.99	29.7	17.9

[a] Region O is not included in nonurban regions because of the size of water demand in relation to state totals. Region O represents 27 percent of all water demand in Texas in 2000, primarily for irrigation, and most of that is from groundwater.

Source: Texas Water Development Board, *Water for Texas, 2007*

Municipal water projections, which include residential and commercial uses, are based on the combined factors of population change and per capita water use. Statewide, the total municipal demand is projected to grow more rapidly than other uses, increasing by two thirds from 2000 to 2040, and more than doubling from 2000 to 2060 (Fig. 7.3). By 2040, water usage is also projected to increase by 50 percent for manufacturing and more than double for steam electric power (Table 7.4).

All regions where Texas urban centers are located are projected to have increased water demand from urban uses: residential, commercial, manufac-

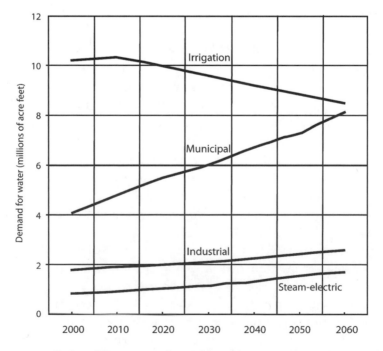

FIGURE 7.3. Projected Texas water demand by major uses, 2000–2060.

TABLE 7.4. Projected water demand by use (in million acre-feet) in Texas, 2000–2040

WATER USE	2000	2010	2020	2030	2040	% CHANGE 2000–2040
Municipal	4.05	4.77	5.48	6.12	6.740	66.5
Manufacturing	1.56	1.83	2.00	2.16	2.32	48.7
Mining	0.28	0.27	0.28	0.29	0.28	−0.9
Steam-electric	0.56	0.76	0.89	1.03	1.17	109.2
Livestock	0.30	0.34	0.37	0.38	0.39	29.2
Irrigation	10.22	10.35	9.98	9.59	9.21	−10.0
Total	16.98	18.31	19.01	19.57	20.10	18.4

Source: Texas Water Development Board, Water for Texas, 2007, p. 123

turing, and steam electric power generation (Table 7.5). Projected demand and water needs (demand minus supply) vary widely among these regions, with projected combined demand increasing by 77 percent (2010–2060). The demand for the remaining regions is projected to increase 50 percent. The increases range from a low of 60 percent in the El Paso region (where considerable water conservation measures are being applied), with irrigation demand being shifted to urban uses. The highest increase (124 percent) is projected for the Lower Rio Grande region. Increases in municipal water needs from 2010 to 2040 are projected to range from almost zero in El Paso (Region E) to a ninefold increase in the Lower Rio Grande (Region M) (Table 7.6).

Although future water resources and demand vary widely from region to region, similar adaptation strategies can be transferred somewhat from area to area. For example, El Paso (a region with limited resources) has already established strict demand-reduction measures while pursuing new technologies for resource recovery and use. In some respects, these areas are already better equipped to address future water supply challenges. Similar approaches will be needed in other Texas cities in response to climate change effects

TABLE 7.5. Projected water demand in Texas regions with major urban centers, 2000–2040

URBAN CENTER	TWDB PLANNING REGION	WATER DEMAND, IN MILLION ACRE-FEET 2000	WATER DEMAND, IN MILLION ACRE-FEET 2040	% INCREASE
Austin	Region K	1.00	1.24	23.2
Dallas–Fort Worth	Region C	1.38	2.62	90.0
El Paso	Region E	0.67	0.70	4.9
Houston	Region H	2.09	2.94	40.9
Lower Rio Grande	Region M	1.33	1.50	12.3
San Antonio	Region L	0.90	1.15	28.8
Total urban regions		7.37	10.15	37.8
Texas		16.98	20.10	18.4

Source: Texas Water Development Board, Water for Texas, 2007

TABLE 7.6. Summary of projected urban water demand and needs in Texas

URBAN AREA (TWDB REGION)	PROJECTED WATER DEMAND
Austin (Region K)	Municipal and manufacturing demand projected to double from 2010 to 2060; steam-electric to increase by almost 50%. By 2040, water needs[a] for municipal use projected to increase seven-fold.
Dallas–Fort Worth (Region C)	From 2010 to 2060, municipal use projected to increase by 92%, steam-electric by 1.5 times, manufacturing by 50%. By 2040, municipal water needs[a] projected to increase 3.7 times.
El Paso (Region E)	Most water used for irrigation. By 2060, municipal use projected to increase by 50%, steam-electric demand to triple. By 2040, municipal water needs[a] projected to increase slightly, with needs of population growth met from existing supplies.
Houston (Region H)	From 2010 to 2060, municipal demand projected to increase by 75%, steam-electric by 138%, manufacturing by almost one third. By 2040, municipal needs[a] projected to increase fivefold.
Lower Rio Grande (Region M)	From 2010 to 2060, municipal demand projected to more than double, steam-electric to increase by 142%. By 2040, municipal needs[a] projected to increase almost ninefold.
San Antonio (Region L)	From 2010 to 2060, municipal demand projected to increase by 50%, with similar increases in manufacturing; steam-electric to more than double. By 2040, municipal needs[a] projected to increase 2.5 times.

[a] Water need is defined as demand minus supply.

Source: Texas Water Development Board, *Water for Texas, 2007*

and to accommodate population growth. Many Texas cities currently use a large portion of their potable water for lawn irrigation and household uses, practices likely to change as the climate changes. Cities can rapidly adopt more stringent, cost-effective water use measures as part of resource planning and decision making. Water demand for future steam electric power can be addressed through energy-efficiency measures (already available to reduce electric power use), peak power demand management, distributed generation technologies, and strategies for renewable energy.

Urban Energy Use

In Texas most electricity and transportation energy use occurs in major cities and urbanized areas. Higher temperatures from climate change will increase total urban electric power demand as well as peak period use. Higher wintertime temperatures will reduce demand for heating in Texas cities, largely from natural gas. The net effect on energy demand will vary from city to city with differences in regional climate.

Cities are also the places where higher levels of efficiency can be achieved to reduce greenhouse gas emissions (e.g., lower fuel use per passenger or ton mile of travel, less distribution loss for electric power due to development densities). This is because of the high population and economic densities in urban areas. For example, with sufficient motivation and policy leadership, all major Texas cities could reduce transportation fuel consumption through increased transit use and other available demand-reduction measures. Likewise, urban development standards can be changed to mitigate the inefficient patterns and styles of development currently dominant in Texas.

Texas electric power systems (generation, transmission, and distribution) are designed with the capacity to meet peak demands during the hottest summer days. Urban air conditioning typically drives electric power demand loads. Peak demand is the critical factor in determining generating capacity and requirements for more robust transmission and distribution systems (U.S. Climate Change Science Program 2007). To adapt to higher urban temperatures (both from climate change and urban heat island effects), the existing building inventory in Texas cities will need to be retrofitted for much higher levels of efficiency, as well as use of distributed generation technologies. This applies particularly to existing buildings that were designed for cheap electricity and lax building standards.

The cooling degree day (CDD) is a measure of electric power demand due to air conditioning. In Texas the number of CDDs varies widely among major urban areas, with 2,094 for El Paso and almost twice that number (4,076) for the Lower Rio Grande. The four other major urban regions range from 2,763 to 3,016 CDDs, but all are more than double the U.S. average of 1,242 CDDs (Table 7.7).

Heating requirements vary even more widely from city to city. The number of heating degree days ranges from 693 in the Lower Rio Grande to 2,708 in El Paso. If climate change resulted in higher wintertime temperatures, Texas would see a reduction in heat energy requirements and in the use of natural gas, the most common source of heating in Texas cities. The cities with the greatest heating demand, El Paso and Dallas–Fort Worth, would experience the most reduction in heating energy use. The adoption of more stringent energy codes in these cities, coupled with extensive retrofitting of

TABLE 7.7. Cooling and heating demand in major Texas cities

CITY	ANNUAL AVERAGE CDD	ANNUAL AVERAGE HDD
Austin–Round Rock	3,016	1,688
Dallas–Fort Worth–Arlington	2,763	2,259
El Paso	2,094	2,708
Houston–Sugar Land–Baytown	3,012	1,371
Lower Rio Grande	4,076	693
San Antonio	2,996	1,644

CDD, Cooling degree days. HDD, Heating degree days.

Source: www.fedstats.gov

184 I The Impact of Global Warming in Texas

the existing building inventory, would further reduce heating energy demand and associated greenhouse gas emissions.

Texas cities are only indirectly involved in electricity generation, distribution, and use, with the exception of those with municipal utilities, such as Austin and San Antonio. Texas cities also play a role in the state's electric utility regulation when they intervene in issues directly affecting them. In 2006, several Texas cities, including Dallas and Houston, joined legal efforts to oppose the permitting of proposed coal power plants in North Texas. Part of their opposition was based on increased carbon dioxide emissions that would be generated by the addition of these plants. The Dallas–Fort Worth region has also faced the challenge of meeting the federal ozone standard, and these new plants were shown to have negative effects on the region's air quality.

City building and development codes also affect electric power consumption. Texas adopted its first statewide building energy code in 2001. The cities of Austin (a national leader in green building standards), Houston, and Dallas have adopted even more stringent building energy codes in recent years. These newly adopted codes expressly exceed the Environmental Protection Agency's Energy Star program. In the 2008 Dallas City Council resolution adopting its green building program, greenhouse gases were specifically referenced as part of the basis for the new program: "Whereas, commercial and residential buildings consume 40 percent of our nation's energy and are responsible for 40 percent of the greenhouse gas emissions in the United States" (Dallas City Council 2008). The tough new standards, however, do not affect existing, less energy efficient buildings, although Austin is developing an ordinance that will require an energy audit as part of the disclosure process for prospective buyers of houses more than 10 years old.

Development codes, such as zoning and subdivision regulations, can allow or require higher-density, more-efficient developments than might be accomplished otherwise. Such development reduces transportation fuel use and associated carbon emissions. Texas cities have not overtly used development codes for the stated purpose of greenhouse gas reductions or mitigation of climate change effects.

Urban Transportation

All modes of transportation are vulnerable to climate change impacts, and in Texas these impacts will vary widely from region to region (Transportation Research Board 2008:84). Transportation modes include land (highways, rail, and pipeline), marine (ports and harbors), and air. Texas cities have infrastructure and operations that will be differentially affected by impacts that include higher temperatures and temperature extremes, heavy precipitation and rising sea level, and more intense tropical storms.

Higher temperatures are primarily a concern for northern U.S. climates (e.g., permafrost melting), but extended periods of high temperature will damage road surfaces, airport runways, bridge joints, and rail lines and disrupt transportation operations. Hotter urban areas are more vulnerable to these types of impacts. Lower water levels in waterways would reduce the capacity of Houston and other Gulf Coast ports. Higher temperatures also affect the lift that is needed for aircraft takeoffs, potentially reducing air travel and cargo capacity (Transportation Research Board 2008:66–67).

Intense precipitation events are already affecting urban transportation in Texas, and these would increase under climate change conditions. During one week in 2006, almost an entire year's worth of rain flooded El Paso, closing streets and damaging roads and drainage infrastructure. In Houston annual heavy rainfall events cause roadway flooding, highway closures, and infrastructure damage. Such events also occur in other Texas cities, producing travel delays and roadway damage. The extensive pipeline networks in Texas, particularly near Houston and the Gulf Coast, can be affected by heavy precipitation events that reduce soil cover and cause subsidence (Transportation Research Board 2008:68). Intense precipitation events can flood port facilities and operations, resulting in shipping delays and interruption of service.

Houston and other Texas Gulf Coast cities are vulnerable to sea-level rise (and land subsidence). The IPCC Fourth Assessment Report on North America identifies this and associated storm surges as "one of the most serious problems" for the Gulf Coast (Transportation Research Board 2008:68). Storm surges, in combination with sea-level rise, can eliminate portions of the Intracoastal Waterway along the Gulf Coast, affecting freight and port movement and ending barge traffic in these areas (Transportation Research Board 2008:69). The IPCC report on human settlements identifies extreme storm events as the principal vulnerability for transportation systems, as well as other impacts, such as sea-level rise and extreme temperature effects on transportation infrastructure (Wilbanks et al. 2007:371).

Hurricanes Katrina and Rita, which hit the Gulf Coast in 2005, are examples of recent storms that disrupted transportation in Texas cities, according to studies by the Transportation Research Board (2008) and the U.S. Climate Change Science Program (Savonis et al. 2008). These events affected all transportation modes to some degree and provided lessons in how transportation systems can be adapted for future storms, including the need for redundancy in transportation systems, the importance of electricity and manpower following such events, and the need for redesign and relocation in anticipation of future storms. Transportation operations during Hurricane Ike in 2008 were considerably improved, possibly because of what was learned from Katrina and Rita, such as staged evacuations and contraflow lanes. Those changes, however, are primarily disaster response measures, and the responses continue to raise issues about the resiliency of urban transportation systems. The implications for future transportation infrastructure include a need for changes in materials, maintenance, and operations, as well as changes in design standards to reflect climate change vulnerabilities.

To date, most studies of climate change and transportation in the United States focus on greenhouse gas reductions, rather than adaptation measures or improvements to the transportation system's resiliency. Similarly, state and federal legislative initiatives (such as the California low carbon fuel standard) have addressed reductions in carbon emissions. Recent regional transportation studies of climate change have identified transportation system vulnerabilities in the study regions, but they do not look at measures to adapt to risks or become more resilient (Savonis et al. 2008). So far, studies of adaptations specific to Texas regions or cities have not been reported.

A 2008 Federal Highway Administration report on climate change and transportation planning includes a chapter on adaptation (ICF International 2008). The report points out that metropolitan planning organizations (MPOs), the regional entities with primary transportation planning responsibilities, and state departments of transportation face large uncertainties with respect to climate change responses. The report notes that MPOs and transportation departments "have little if any information on precisely what impacts they can expect, where, and in what time frames. As a result, agencies are largely not acting to adapt the transportation system to climate change, or are waiting for further guidance on the topic" (ICF International 2008:31). Texas MPOs and the transportation department are likely to await guidance from federal agencies before incorporating climate change in the planning process. Additionally, they may choose to learn from the recent experience of other states that are more actively engaged in climate change policies and studies (e.g., Washington, New York, Florida, and California).

Long-range transportation plans in Texas do not address climate change impacts or vulnerabilities. Only Houston's long-range plan mentions climate change as part of the context for long-range planning. None of the plans discuss carbon issues or associated issues that arise from escalating fuel and energy prices that directly affect infrastructure financing (Capital Area MPO 2005; El Paso MPO 2007; North Central Texas Council of Governments

2007; San Antonio–Bexar County MPO 2004). Indirect references to carbon emissions include more efficient transportation systems (i.e., transit, bicycle, ride-sharing options, better fuel mileage) and goals for increasing use of "clean, alternative fuels," which in Texas has included such fossil fuels as natural gas propane and low-sulfur diesel. None of the plans include explicit goals or actions to reduce transportation carbon emissions as part of climate change concerns or to adapt transportation to climate change impacts.

The plan for the Houston area released in 2007 includes a brief discussion of climate change and transportation. The region's MPO participated in recent climate change studies of the Gulf Coast region and has initiated a process to examine climate change impacts (organized under the title "Foresight Panel for Environmental Effects"). The regional transportation plan states that inclusion of climate change is a first step in further consideration of the effects on regional transportation. Although the plan points out that the region's air-quality actions may reduce carbon emissions, no specific policies or adaptation measures are discussed. The inclusion of climate change, remarkable in and of itself, may have been a response to the extraordinary climate events and conditions in this region, which has been marked by recent tropical storms and frequent severe flooding.

According to the Federal Highway Administration (FHWA), many states and MPOs are beginning to include climate change issues in transportation planning. The FHWA also reports that many of these organizations believe that effective measures to reduce greenhouse gas emissions and to adapt to vulnerabilities are largely outside their responsibilities and capacities (ICF International 2008:36). With transportation responsibilities divided across many entities (federal, state, and local), the ability of any particular entity to respond is restricted.

Reducing greenhouse gas emissions from transportation is a relatively straightforward process if emissions are treated in a manner similar to current air-quality planning. This approach involves difficult technical and regulatory issues, but an emission-reduction context parallels much of what occurs in air-quality planning in Texas. National transportation researchers and other state air-quality organizations have been identifying and quantifying carbon emission reduction strategies for several years, and many of the analytical and modeling tools are already available to Texas.

Adaptation to climate change effects is not likely to be part of any single known process at this point in time. Adaptive responses for transportation systems in major Texas cities would vary widely with respect to vulnerabilities and local needs. Houston and Gulf Coast cities will need to focus on more severe and frequent storms, disaster response, and flooding. This means redesign and relocation of parts of the transportation infrastructure in ways that provide system redundancy and resilience. Public transportation, railroads, freight, pipelines, ports, and air travel are essential components of regional transportation that must be part of adaptation measures in Texas cities. New ways of planning, managing, and financing such responses will be needed.

Adaptation measures in other major Texas cities must incorporate higher temperatures, extended heat waves, and changes in precipitation levels in shaping future transportation systems. These impacts can affect the facility design and the materials used in transportation infrastructure, as well as maintenance of the infrastructure. All of these adaptations require strategic planning, operational planning, and adequate financing.

Human Health
Air Quality

Higher temperatures will result in worsened air quality in Texas urban areas, including longer ozone seasons and higher summer ozone concentrations. Higher temperatures can also increase the potential for higher concentrations of health-damaging fine particulates. In the Dallas–Fort Worth area, a longer ozone season would accompany higher temperatures, extending into the spring and fall months. In the Houston area, higher winter temperatures would ensure a year-round ozone season. Higher temperatures also produce higher levels of biogenic volatile organic compounds (VOCs), such as isoprene, which would contribute both to higher ozone concentrations and to higher levels of background ozone that can move into urban airsheds. For major Texas cities, except possibly El Paso, oxides of nitrogen (NOx) will need to be reduced even more drastically, perhaps approaching zero for some sources. As a primary precursor to ozone formation, this poses a particularly difficult challenge. NOx emissions are a product of all hydrocarbon combustion, whether from gasoline or diesel engines, coal or natural gas in power plants, or fossil fuels in industrial boilers.

Ozone, a reactive form of oxygen (O_3) that can damage human respiratory tissue as well as vegetation, is regulated as a criteria pollutant under federal law. It forms in the presence of sunlight through chemical reactions involving two primary precursors, NOx (primarily from fuel combustion from industry, motor vehicles, and power plants) and VOCs (from evaporation, industrial processes, and biogenic sources). Heat and sunlight are an essential part of these chemical reactions (Walcek and Yuan 1995).

In May 2008 the federal 8-hour standard for ozone was exceeded in Houston (and eight nearby counties), Dallas–Fort Worth (with nine counties), and Beaumont–Port Arthur (three counties). El Paso exceeded federal standards for carbon monoxide and particulate matter. Austin, San Antonio, and Northeast Texas are early action compact areas that have submitted plans intended to keep ozone levels lower than the federal standard. The Houston and Beaumont–Port Arthur metropolitan areas have requested more stringent designations (severe and moderate, respectively), because they have been unable to demonstrate achievement of the current standard. In 2010, the Environmental Protection Agency proposed strengthening the 8-hour standard from 0.08 to within the range 0.06–0.07 parts per million. Several additional Texas urban areas are expected to be in nonattainment as a result.

Increased temperatures in cities, whether they are induced by climate changes or urban heat island effects, intensify ozone formation. For Los Angeles, every 1.8°F increase in temperature above 71.6°F could increase ambient ozone concentration by 5 percent (Akbari 2005). Similar increases can be expected in Texas cities. Higher temperatures also require more energy for air conditioning, thereby providing additional ozone precursors (NOx) from power plant emissions. Higher temperatures increase the rate of evaporation of VOCs from vehicle fuels (i.e., hydrocarbons in gasoline and additives) and from biogenic sources released by some plant species in response to higher temperatures (i.e., isoprene from various species of oak trees).

Heat Mortality

Higher temperatures, particularly if they occur as extended heat waves, negatively affect human health. Heat mortality is the largest weather-related cause of death in the United States (Davis et al. 2003), although death and mortality are underreported, since victims usually have other contributing health conditions (Wilhelmi et al. 2004). Extended heat events occur when temperatures remain above a threshold level for several days. The thresholds, which include temperature and humidity, are higher in cities with hotter climates because they have adapted to local conditions; for example, the threshold is 110°F in Houston, but it drops to 103°F in Chicago and only 88°F in Portland (Kalkstein and Greene 2007). Heat-related illnesses and deaths have declined as the availability of air conditioning has grown. Nevertheless, extreme heat waves in northern U.S. cities and Europe raise longer-term concerns that heat events will overload electric power systems and disrupt air-conditioning availability. Although the Texas population is more adapted to higher temperatures, vulnerable urban populations, including anyone lacking access to air-conditioned space, will need protection from extended heat waves.

Infectious Diseases

Increased temperatures and changes in rainfall from climate change contribute to the occurrence of disease and its transmission across the human population. Cities are particularly vulnerable because of high population concentrations and rapid migration into cities during times of stress. Projections by researchers generally support the view that health effects from climate change will be negative (Institute of Medicine 2008). The diseases discussed in this context include malaria, dengue fever, tick-borne diseases, and diseases associated with diarrhea, such as cholera (Institute of Medicine 2008). Malaria is virtually nonexistent in the United States, but dengue fever infected almost 4,000 people in the last 25 years. In 2002, West Nile virus, also carried by mosquitoes, was identified in bird populations in Houston. The virus was first detected in New York in 1999, and in 2008 human infections were reported in 11 Texas counties (Texas Department of State Health Services 2008). Following Hurricane Katrina, there was concern in the Houston area that infectious diseases would be brought from Louisiana. There were a few cases of vibrio vulnificus, a noncontagious virus in the cholera family, but it was contracted from direct exposure to contaminated water, not exposure to people. Speculations about post-storm disease migration to Houston have not been realized.

IPCC reports and studies indicate that the severity of disease effects is expected to be greatest worldwide in low-lying coastal areas that lack the capacity to respond effectively to health needs. This suggests that "capacity to respond" is a key factor in considering health effects, rather than disease-specific factors. Currently, large U.S. cities, including those in Texas, have the best capacity to address medical and public health responses. Unfortunately, it cannot be known with any degree of certainty whether these capacities will be sufficient in the decades ahead under alternative climate scenarios.

News reports of possible disease outbreaks often identify alternative causes, with climate change being mentioned whether there is evidence of a causal link or not. Another unknown to consider for future effects is the interplay of societal factors and disease exposure. For example, availability of air conditioning across most income levels in major Texas cities has greatly reduced exposure to mosquito-borne diseases. Under a favorable scenario, large improvements in energy efficiency, use of renewable energy, and more resilient infrastructure could further the trend toward affordable, sustainable air-conditioned space. Alternatively, under conditions of economic and energy failures, many more people could be exposed to mosquito-borne diseases (regardless of whether this exposure is driven by climate change effects). Likewise, the future availability of effective disease prevention and treatment or of measures to control mosquitoes is unknown or highly uncertain in the time frame of climate change effects.

Within this framework of uncertainty, health system responses to vulnerabilities are being considered (Institute of Medicine 2008). First and foremost is the need for effective long-term monitoring of disease dynamics associated with climate change. Without such information, it will be difficult to distinguish between a single outbreak and more widespread problems. At the same time, the capacity to respond to such diseases must be present as part of the state's public health system and, more important, within major cities where vulnerable populations are concentrated. The training of health professionals working in Texas will need to include detection and treatment of diseases that are more likely to occur under changed climate conditions. These too may vary regionally across the state.

URBAN LEADERSHIP

While there has been relatively little response in Texas to climate change concerns, the leadership that has emerged is found among local government officials. Several Texas mayors are participating in climate-change initiatives, and 22 have signed the Conference of Mayors Climate Protection Agreement, including the mayors of Austin, Dallas, San Antonio, Fort Worth, El Paso, and Arlington. The city of Arlington, a suburban community of 370,000 located between Dallas and Fort Worth, has published a baseline inventory of greenhouse gas emissions including municipal operations, which are controlled by local government, and the community at large (City of Arlington 2008). Arlington's mayor, Robert Cluck, has played a prominent role in state and national climate discussions. In 2008, the mayor of Houston, Bill White, issued a plan for substantially reducing city government carbon emissions (those generated by city functions) by 2010. The city of Austin's Climate Protection Plan, published in 2007 with leadership from Austin's current and previous mayors, sets forth the goal of making all city facilities, vehicles, and operations carbon neutral by 2020. Other plan components address the city's electric utility (Austin Energy), homes and buildings, and the broader community's carbon emissions. The current mayors of Houston, Arlington, and Austin have been active in efforts to acknowledge climate issues and to reduce related carbon emissions. Austin, Houston, and San Antonio are also designated solar cities under a current U.S. Department of Energy initiative.

To date, urban leadership in response to climate change has been primarily about reducing carbon emissions through energy efficiency, transportation measures, and alternative energy. Severe storms, flooding, and sea-level rise have been viewed largely in the context of disaster response and storm-water management, although citizens often link climate change with these events. The longer view of adaptation and resilience is only now becoming part of the vocabulary of urban issues as the reality of climate change becomes an accepted part of the public dialogue.

CONCLUSIONS

As population and economic centers, Texas cities are particularly vulnerable to the impacts of global climate change. These vulnerabilities vary from city to city, depending on their location with respect to climate change risks and their subregional climate conditions. Coastal population centers, from Houston to the Lower Rio Grande Valley, are vulnerable to sea-level rise, increased storm intensity, and accompanying flooding. The impacts can affect millions of people while disrupting and damaging road, air, and port transportation systems. All major Texas cities face the possibility of impacts on air quality, energy, health, and other temperature-related effects. All major cities face the prospect of declining water resources within the time frame examined here.

Cities and urban regions that currently fail to meet air-quality health standards will likely continue to struggle with these standards as ambient temperatures climb and urban heat island effects intensify. Background ozone levels are already high enough to trigger air-pollution events in major Texas cities. If current state and federal pathways are followed in the absence of broader changes, urban air-quality goals are unlikely to be accomplished. In Texas, the emerging attention to carbon emissions could serve as a key element for pursuing both improved air quality and greenhouse gas reductions in concert. Known and foreseeable changes in transportation (including fuels, vehicles, and travel modes) could reduce transportation emissions significantly, even approaching zero net carbon levels in some sectors. This would require redirection of state and national actions to restructure urban transportation and energy policies, as well as changes in travel and related consumer behaviors. Both air quality and greenhouse gas levels will benefit from rapid movement toward a more electric transportation system and the use of low carbon Texas renewable energy sources.

Major cities are vulnerable to higher electric power demand and insufficient water resources, but there are known ways of reducing those vulnerabilities. Existing and near-term technologies can improve the efficiencies of electric power and water management systems—wasteful lawn irrigation practices being a well-known example of needed water resource redirection. Energy consumption in buildings could be cut in half or more through efficiency improvements and to net-zero carbon through widespread deployment of distributed generation technologies for buildings. Such technologies are available and relatively well known, but more responsive governmental, economic, and consumer actions will be needed for cities to make these changes (Harriss 2007).

Texas cities have abilities and capacities to adapt and otherwise respond to the consequences of climate change, thanks to their relative wealth, infrastructure, financing capabilities, health systems, and governance capabilities. Some cities may incur gradual climate change effects (e.g., rising nighttime temperatures or declining water availability), while others will experience abrupt events, such as tropical storms or severe flooding. The inertia of existing urban investments (such as residential patterns, water and sewer services, transportation systems, and electric power systems) poses difficult and expensive barriers to the retrofitting of cities. Foresight, leadership, and financing are needed for some adaptations that will require decades to achieve. Decisions on such changes are needed despite the uncertainties about future vulnerabilities.

Cities need common guidelines and regional decision frameworks to help them consider how best to respond. Changes in governance authority will likely be needed as well. Hundreds of communities within urban areas need to respond in ways that are somewhat consistent with each other and with the vulnerabilities facing them. Regional transportation planning will change at some point in response to federal guidelines and funding. Water resource planning will change in response to state guidelines, drawing from best practices in other areas of the country. Utilities will react to price changes and to federal regulatory changes in response to climate effects. Public health services as well as disaster preparedness functions will change in response to conditions and events. All of these functions need information that is garnered from research, data assembly, program design and testing, and policy interactions among the affected entities, including frequent interaction with decision makers and stakeholders.

REFERENCES

Akbari, Hashem, 2005. *Energy Saving Potentials and Air Quality Benefits of Urban Heat Island Mitigation.* LBNL-58285, First International Conference on Passive and Low Energy Cooling for the Built Environment, Athens, Greece. http://www.osti.gov/energycitations/product.biblio.jsp?osti_id=860475.

Capital Area MPO, 2005. *CAMPO Mobility 2030 Plan.* Capital Area Metropolitan Planning Organization, Austin, Tex.

Christensen, J. H., B. Hewitson, A. Busuioc, A. Chen, X. Gao, I. Held, R. Jones, R. K. Kolli, W. T. Kwon, R. Laprise, V. Magaa Rueda, L. Mearns, C. G. Menndez, J. Risnen, A. Rinke, A. Sarr, and P. Whetton, 2007. Regional Climate Projections. Pp. 847–940 in: *Climate Change 2007: The Physical Science Basis*. Contribution of Working Group 1 to the Fourth Assessment Report of the Intergovernmental Panel on Climate Change. Cambridge University Press, Cambridge, U.K., and New York.

City of Arlington, 2008. *Greenhouse Gas Emissions Inventory*, August 2008. City of Arlington, Tex.

Dallas City Council, 2008. Revised Agenda Item 28, April 9, 2008. City of Dallas, Tex.

Darby, L., and C. J. Senff, 2007. Comparison of the Urban Heat Island Signatures of Two Texas Cities: Dallas and Houston. Seventh Symposium on the Urban Environment, American Meteorological Society, Joint Session J2, 10-13 September 2007, San Diego, Calif. (Characterizing the Urban and Coastal Climate: Thermal and Boundary Layer Structure and Atmospheric Responses). http://ams.confex .com/ams/7Coastal7Urban/techprogram/session_20893.htm.

Davis, R., P. Knappenberger, P. Michaels, and W. Novicoff, 2003. Changing Heat-Related Mortality in the United States. *Environmental Health Perspectives* 111(14):1712.

El Paso MPO, 2007. *Trans-Border 2035 Metropolitan Transportation Plan*. El Paso Metropolitan Planning Organization, El Paso, Tex.

Environmental Protection Agency, 2008. *Urban Heat Island Reduction Initiative (HIRI)*. Environmental Protection Agency, Washington, D.C. http://www.epa .gov/hiri/index.htm.

Harriss, R., 2007. An Ongoing Dialogue on Climate Change: The Boulder Manifesto. Pp. 485–490 in: *Creating a Climate for Change*. Susanne C. Moser and Lisa Dilling (eds.). Cambridge University Press, New York.

Hitchcock, D., 2004. *Cool Houston Plan*. Houston Advanced Research Center, Houston, Tex.

ICF International, 2008. *Integrating Climate Change into the Transportation Planning Process*. Final Report, July 2008. Federal Highway Administration, Washington, D.C.

Institute of Medicine, 2008. *Global Climate Change and Extreme Weather Events: Understanding the Contributions to Infectious Disease Emergence*. Rapporteurs: D. A. Relman, Margaret A. Hamburg, Eileen R. Choffnes, and Alison Mack. National Academies Press, Washington, D.C. http://www.nap.edu/catalog/12435 .html.

Kalkstein, L., and J. S. Greene, 2007. An Analysis of Potential Heat-Related Mortality Increases in U.S. Cities under a Business-as-Usual Climate Change Scenario. Manuscript.

North Central Texas Council of Governments, 2007. *Mobility 2030*. North Central Texas Council of Governments, Dallas–Fort Worth Region.

Oke, T. R., 1973. City Size and the Urban Heat Island. *Atmospheric Environment* 7:769–779.

Rose, S. L., H. Akbar, and H. Taha, 2003. *Characterizing the Fabric of the Urban Environment: A Case Study of Metropolitan Houston, Texas.* LBNL-51448. Lawrence Berkeley National Laboratory, Berkeley, Calif.

Ruth, M., and P. Kirshen, 2002. Dynamic Investigations into Climate Change Impacts on Urban Infrastructure: Background, Examples, and Lessons. Presented at Western Regional Science Association Annual Meeting, Monterey, Calif.

San Antonio–Bexar County MPO, 2004. *Mobility 2030, Metropolitan Transportation Plan.* San Antonio–Bexar County Metropolitan Planning Organization, San Antonio, Tex.

Savonis, M. J., V. R. Burkett, and J. R. Potter, 2008. *The Impacts of Climate Change and Variability on Transportation Systems and Infrastructure: Gulf Coast Study.* Phase I, U.S. Climate Change Science Program Synthesis and Assessment Product 4.7. U.S. Department of Transportation, Washington, D.C.

Stone, B., 2007. Urban and Rural Temperature Trends in Proximity to Large U.S. Cities, 1951–2000. *International Journal of Climatology* 27:1801–1807.

Streutker, D. R., 2002. A Remote Sensing Study of the Urban Heat Island of Houston, Texas. *International Journal of Remote Sensing* 23(13):2595–2608.

Texas Department of State Health Services, 2008. http://www.dshs.state.tx.us /idcu/disease/arboviral/westNile/maps/texas/2007/txHumanWNF.jpg, accessed November 2008.

Texas Water Development Board, 2007. *Water for Texas, 2007.* Texas Water Development Board, Austin.

Transportation Research Board, 2008. *Potential Impacts of Climate Change on U.S. Transportation.* Special Report 290, Committee on Climate Change and U.S. Transportation, Transportation Research Board. National Research Council, Washington, D.C.

U.S. Climate Change Science Program, 2007. Effects of Climate Change on Energy Production and Use in the United States. U.S. Climate Change Science Program, Subcommittee on Global Change Research, Washington, D.C. P. 102.

Walcek, C. J., and H. Yuan, 1995. Calculated Influence of Temperature-Related Factors on Ozone Formation Rates in the Lower Troposphere. *American Meteorological Society* 34:1056–1069.

Wilbanks, T. J., P. Romero Lankao, M. Bao, F. Berkhout, S. Cairncross, J.-P. Ceron, M. Kapshe, R. Muir-Wood, and R. Zapata-Marti, 2007. Industry, Settlement, and Society. Pp. 357–390 in: *Climate Change 2007: Impacts, Adaptation, and Vulnerability.* Contribution of Working Group 2 to the Fourth Assessment Report of the Intergovernmental Panel on Climate Change. Cambridge University Press, Cambridge, U.K., and New York.

Wilhelmi, O., K. Purvis, and R. Harriss, 2004. Designing a Geospatial Information Infrastructure for Mitigation of Heat Wave Hazards in Urban Areas. *Natural Hazards Review*, ASCE, 5(3):147–158.

CHAPTER 8

Greenhouse Gas Emissions
Judith Clarkson

As outlined in Chapter 1, the warming of the earth is directly related to the atmospheric concentration of greenhouse gases, notably carbon dioxide (CO_2). Since preindustrial times, greenhouse gas emissions have grown significantly, with an increase of 70 percent between 1970 and 2004. Emissions of the various gases have increased at different rates, with CO_2 emissions growing about 80 percent between 1970 and 2004 and representing 77 percent of total greenhouse gas emissions in 2004. This has resulted in an increase in atmospheric CO_2 from a preindustrial value of about 280 parts per million (ppm) to 379 ppm in 2005. This exceeds by far the natural range over the last 650,000 years (180-300 ppm), as determined from ice cores. Over the same period, methane emissions have increased by 148 percent (IPCC 2007a).

The largest growth in global greenhouse gas emissions between 1970 and 2004 has come from the energy supply sector (an increase of 145 percent). The growth in direct emissions from transportation was 120 percent; industry 65 percent; and land use, land use change, and forestry 40 percent. The most important factors in these increases are global income growth (77 percent) and global population growth (69 percent). These increases have more than offset a decrease in global energy intensity of 33 percent between 1970 and 2004, although the long-term trend of a declining carbon intensity of energy supply reversed after 2000 (IPCC 2007c).

Per capita emissions of CO_2 vary enormously, with 20 percent of the world population producing 57 percent of world gross domestic product and accounting for 46 percent of global greenhouse gas emissions. As the rest of the world strives to attain a higher standard of living, emissions are expected to increase dramatically, with an increase in warming of about 0.2°C (0.36°F) per decade projected for a range of emission scenarios. Carbon dioxide emissions from energy use between 2000 and 2030 are projected to grow 40–110 percent. Two thirds to three quarters of this increase is expected to come from developing regions (IPCC 2007c).

An increase in global average surface temperature following a doubling of CO_2 concentrations is likely to be in the range of 2–4.5°C (3.6–8°F), with a best estimate of about 3°C (5.4°F); values substantially higher than 4.5°C cannot be excluded (IPCC 2007a). According to the IPCC (2007b), a change in global temperatures of this magnitude could result in "abrupt and irreversible" impacts that could include the fast melting of glaciers and species extinctions. As many as 20–30 percent of species assessed so far are likely to be at risk if increases in global average temperatures exceed 1.5–2.5°C, relative to the 1980–1999 average. Even if levels of CO_2 in the atmosphere stay where they are now, it is likely that temperatures will continue to increase about 0.1°C per decade because of the slow response of the oceans. Without extra measures, CO_2 emissions will continue to rise; they are already growing faster than a decade ago, partly because of increasing use of coal. The IPCC's economic analyses indicate that the trend can be reversed at reasonable cost. Indeed, the 2007 report says, there is "much evidence that mitigation actions can result in near-term co-benefits (e.g., improved health due to reduced air pollution)" that may offset some costs (IPCC 2007c). This reversal needs to come within a decade or so if the worst effects of global warming are to be avoided.

It is generally agreed that the richest countries need to reduce their greenhouse gas emissions 60–80 percent. Such a reduction could totally change our current lifestyles. The developed world has a very energy-intensive way of life, and this is particularly true of the United States, which uses approximately twice the amount of energy per unit of gross domestic product that Japan and Europe use. This discrepancy alone illustrates that significant opportunities exist for reducing the amount of energy consumed, without a significant deterioration in living standards. In response to the 1973 oil embargo, some attempts to reduce energy consumption were initiated by the U.S. government, but the return of low energy prices in the 1980s reduced a lot of the motivation. An unwillingness to address energy consumption and global warming was particularly evident after the election of President Bush in 2000. He initially said that he would address the issue, but then backed down under pressure from the energy industry.

Overall, the federal government has made little effort to address the issue of global warming, and in general, policies that would reduce emissions are primarily aimed at reducing energy use and our dependence on imported oil. With the election of a Democratic Congress in 2006, there was some movement on the issue. An energy bill enacted in December 2007 increases the fuel efficiency standard of cars to 35 miles per gallon (mpg) by 2020, among other provisions. Other important measures, however, including those to increase renewable energy production, were omitted.

With the election of President Barack Obama, there is a promise of real change at the federal level. Reducing greenhouse gas emissions is one of his legislative priorities, and the House of Representatives has already passed an energy bill (H.R. 2454) that would reduce these emissions by 17 percent by 2020 (83 percent by 2050). Its provisions include requiring utilities to generate an increasing amount of power from renewable sources and a cap-and-trade program. So far the bill is stalled in the Senate. Meanwhile, the Environmental Protection Agency (EPA), using its authority under the Clean Air Act, is moving ahead with regulations to limit greenhouse gas emissions. In addition, in response to the recession, President Obama initiated legislation (the American Recovery and Reinvestment Act of 2009) that allocated more than $60 billion to energy-related projects, including energy efficiency, renewable energy, intercity passenger rail, and carbon capture experiments, as well as funds to modernize the nation's electrical grid.

In the past, most of the implementation and development of new initiatives has taken place at the state level. In this chapter we examine the potential problems and opportunities for Texas as we move toward an era of increasing pressures, both economic and regulatory, to improve energy-use efficiency and limit greenhouse gas emissions. Some states have already adopted programs aimed at increased energy-use efficiency, and an analysis of some of these programs is included.

THE CONTRIBUTION OF GREENHOUSE GAS EMISSIONS IN TEXAS TO GLOBAL WARMING
The Nature of Greenhouse Gases

Earth's atmosphere is 99 percent oxygen and nitrogen; the other 1 percent is made up of a number of trace or greenhouse gases. These gases include carbon dioxide, carbon monoxide, methane, nitrous oxide, ozone, and, for the last 50 years, chlorofluorocarbons (CFCs). The majority of shortwave solar radiation passes through the atmosphere and is absorbed by the earth's surface. The longwave radiation emitted by the earth's surface interacts with the gases in the atmosphere, resulting in the absorption of some and the re-emission of the balance. As trace gases increase in relative amount, the quantity of heat retained by the atmosphere increases.

The most prominent of the greenhouse gases is CO_2. In nature, CO_2 is part of an integrated carbon cycle, in which plants absorb CO_2 from the atmosphere, storing the carbon and releasing oxygen. In turn, CO_2 is released back into the atmosphere as a result of respiration, organic decay, combustion, and chemical diffusion. A steady state results if these processes are in equilibrium. Human activity, however, has caused an acceleration of many of these processes, resulting in a disproportionate increase in the production of CO_2. One of the more significant causes is the rapid release of CO_2 from fossil fuels, which have acted as a carbon sink over a very long period of time. A second major contributor is the destruction of forests, which not only release their existing carbon but also no longer have the capacity to contribute to the carbon cycle by removing CO_2 from the atmosphere.

Methane is a naturally occurring gas that makes up approximately 18 percent of the greenhouse gas currently contributed by human activity. It is generated naturally by bacteria that break down organic matter in the absence of oxygen. Methane may be produced by natural wetlands, flooded rice fields, ruminant livestock, landfills, coal mining, biomass burning, and deforestation. The melting of Arctic permafrost as a result of higher temperatures also releases methane. The concentration of methane has more than doubled since the industrial revolution (Graedel and Crutzen 1989).

Recognizing that stratospheric ozone was being destroyed by CFCs, an international agreement, the 1987 Montreal Protocol, was signed by representatives of more than 40 industrialized countries. It aimed to phase out CFC production. By 2004 the emissions of these gases were about 20 percent of their 1990 level (IPCC 2007c). In contrast, the concentration of ozone in the lower atmosphere is increasing. Here it acts as an infrared absorber, contributing to global warming. It is produced mainly as a result of emissions from fossil fuel combustion and is the main component of photochemical smog.

Although industrialized countries were able to come to an agreement to phase out CFCs, efforts to control greenhouse gases have been far less successful. Countries such as the United States, Canada, and Australia, in particular, have resisted any efforts that they perceive will impede their economic growth. After two and a half years of intense negotiations, the Kyoto Protocol was adopted at the third conference of the parties to the Framework Convention on Climate Change in Kyoto, Japan, in December 1997. Most of the world's countries eventually agreed to the protocol, but some nations, including the United States, chose not to ratify it. Following ratification by Russia, the Kyoto Protocol entered into force in February 2005. It requires each of the developed countries to reduce its greenhouse gas emissions below a specified level by 2012. This will result in a total cut in greenhouse gas emissions of at least 5 percent against the baseline of 1990 (Finn 2004). The protocol places a heavier burden on developed nations under the principle of "common but differentiated responsibilities." Developed countries can more easily pay the cost of cutting emissions and have historically contributed more to the problem by emitting larger amounts of greenhouse gas per person.

With the period covered by the Kyoto Protocol nearing conclusion, a follow-up agreement was deemed necessary. As more and more countries have come to understand the urgency of the situation, a more concerted effort to address the issue of climate change that would include participation by developing countries has been initiated. In December 2007 the United Nations climate change convention, on the Indonesian island of Bali, concluded with an agreement to launch a two-year negotiating process, the "Bali roadmap," aimed at securing a binding deal at the 2009 U.N. summit in Denmark. The agreement acknowledged that evidence for the planet's warming is "unequivocal" and that delays in reducing emissions increase the risks of "severe climate change impacts." Attempts by the European Union, China, and many developing countries to get the richest countries to agree to cut their emissions by 25–40 percent were vigorously rejected by the United States (Vidal et al. 2007).

By 2009 the United States had an administration that was committed to a global attempt to reduce greenhouse gas emissions. The agreement reached at Copenhagen, however, was a weak outline, falling far short of what many countries were seeking. After eight draft texts and all-day talks between 115 world leaders, it was left to Barack Obama and Wen Jiabao, the Chinese premier, to broker a political agreement. The so-called Copenhagen accord "recognizes" the scientific case for keeping temperature rises to no more than 2°C but does not contain commitments to reduce emissions to achieve that goal (Vidal et al. 2009).

Greenhouse Gas Emissions in Texas

According to data compiled by the California Energy Commission from a variety of sources, only 12 states had more greenhouse gas emissions per unit of gross state product than Texas in 2001. While Texas had approximately 900 metric tons per million dollars of gross state product, the three highest emitters exceeded 1,500 tons. All three states are coal-mining states, with Wyoming, the highest, emitting more than 3,000 tons. The four states with the lowest levels, including California, emitted approximately 300 tons, and the U.S. average was 600 tons. Internationally, California had emission levels comparable to those of Japan, Italy, and France. Texas was closest in emissions to Australia. Some of the highest levels were emitted by India (2,200 tons), China (3,000 tons), and Russia (5,000 tons). When these emissions are expressed as metric tons per capita, however, the United States has the highest level, 8 times that of China and 20 times that of India (California Energy Commission 2006).

With its large population and energy-intensive economy, Texas leads the nation in energy consumption, accounting for more than one tenth of total U.S. energy use. Energy-intensive industries in Texas include aluminum, chemicals, forest products, glass, and petroleum refining. Texas' petroleum refineries can process more than 4.6 million barrels of crude oil per day, and they account for more than a quarter of total U.S. refining capacity. Most of the state's refineries are clustered near major ports along the Gulf Coast, including Houston, Port Arthur, and Corpus Christi. These coastal refineries have access to local Texas production, foreign imports, and oil produced offshore in the Gulf of Mexico, as well as the U.S. government's Strategic Petroleum Reserve.

Texas crude oil reserves represent almost a quarter of total U.S. oil reserves, and Texas natural gas reserves account for almost 30 percent of the U.S. total. The nation's leading natural gas producer, Texas accounts for more than a quarter of total U.S. gas production, with 10 natural gas marketing centers (more than any other state) and a natural gas storage capacity among the highest in the nation. Most of these active storage facilities are depleted oil and gas fields that allow Texas to store its natural gas production during the summer, when national demand is typically lower, and to ramp up delivery quickly during the winter months when markets across the country require greater volumes of natural gas to meet home heating needs (Energy Information Administration 2008a).

The source of CO_2 emissions in Texas is characterized here by use and fuel in 2005 (Table 8.1). With the exception of cement manufacturing, all of these emissions are a result of combustion of coal (32 percent of total), natural gas (30 percent), or petroleum products (37 percent). Energy consumption data by source and year were obtained from the Energy Information Administration (2008b), and these values were used to determine quantities of CO_2 using conversion factors provided by the Environmental Protection Agency (EPA 2007). Emissions for each fuel were then assigned to a use category, using data from the Energy Information Administration (2008b). Electricity generation produces the most greenhouse gas emissions, at 43 percent of the Texas total, followed by transportation (30 percent) and industrial uses (23 percent).

Power plants fired by natural gas typically account for about half of the electricity produced in Texas, and coal-fired plants account for much of the remaining generation. Texas consumes more coal than any other state, and its emissions of carbon dioxide and sulfur dioxide are among the highest in the nation. Almost all of the emissions produced from coal are a result

TABLE 8.1. Carbon dioxide emissions (in million short tons) in Texas for 2005, by sector

SOURCE	TEXAS TOTAL	SECTOR					% U.S. TOTAL
		TRANSPORTATION	INDUSTRIAL	ELECTRICITY	RESIDENTIAL	COMMERCIAL	
Coal	221.91		9.56	212.35			9.35
Natural gas	212.21	4.98	98.22	88.26	11.14	9.62	16.02
Distillate	60.06	49.26	9.41	0.12		1.27	8.51
Jet fuel	35.61	35.61					13.11
LPG	34.88	0.04	33.94		0.76	0.13	55.80
Motor gas	113.46	111.07	2.35			0.07	8.33
Residual fuel	14.21	12.24	0.23	1.75			7.75
Other petroleum products	2.11		2.11				
Cement manufacturing[a]	5.16		5.16				
Total	699.62	213.19	160.98	302.47	11.90	11.10	10.99

[a] Based on Texas Natural Resource Conservation Commission report (TNRCC 2002), assuming cement manufacturing increased by 2 percent per year since 1999

Data source: Based on Energy Consumption Estimates for Texas (Energy Information Administration 2008b), except for cement manufacturing

of electricity generation, accounting for 70 percent of emissions from the generation of electricity. Natural gas accounts for 29 percent, but because it produces more than twice the energy for the same amount of CO_2 produced, this is not a true reflection of the importance of natural gas for electricity production. Natural gas is also an important input for manufacturing; total natural gas use in Texas accounts for 16 percent of total U.S. usage.

In Texas, per capita residential use of electricity is significantly higher than the national average, because of high demand for electric air-conditioning during hot summer months. Petroleum consumption is the highest in the nation, and the state leads the country in consumption of asphalt and road oil, distillate fuel oil, jet fuel, liquefied petroleum gases (LPG), and lubricants. Texas LPG use is greater than the LPG consumption of all other states combined, primarily because of the state's active petrochemical industry, which is the largest in the United States (Energy Information Administration 2008a). Almost half of the emissions from petroleum products result from gasoline consumption in motor vehicles.

Between 1990 and 2005, the state increased its emissions overall by 13 percent, although total emissions have decreased by 4.6 percent since 2000 (Fig. 8.1). When adjusted for population growth, per capita emissions remained almost constant during the 1990s and then declined by 13 percent between 2000 and 2005. As discussed in Chapter 9, this is a result of the increasing importance of the service sector to the economy. Natural gas use increased 18.6 percent during the 1990s and then declined to slightly below

1990 levels between 2000 and 2005, probably as a result of price increases that reflect a reduction in domestic supply and the difficulties associated with importing natural gas.

Emissions from petroleum products have increased 30 percent since 1990. For motor gasoline, this figure was 36 percent, somewhat higher than the increase in population. Statewide per capita vehicle miles traveled (VMT) increased 10 percent between 1990 and 2000, but the increase was 8 percent for the 1990–2005 period. In the major urban areas, per capita VMT increased throughout the 1990s, and this increase was greatest between 1992 and 1995, particularly in Austin (up 25 percent) and Dallas–Fort Worth (up 20 percent). Since 2000, per capita VMT has remained relatively constant or has declined. For the period 1992–2005, per capita VMT increased 19 percent for San Antonio, 16 percent for Austin, 14 percent for Dallas–

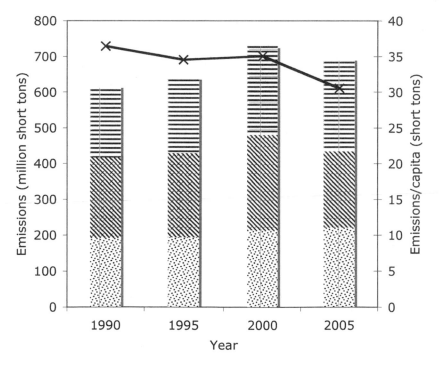

FIGURE 8.1. Sources of carbon dioxide emissions in Texas, 1990–2005. Bars represent emissions of carbon dioxide from coal, natural gas, and petroleum sources, in millions of short tons, as indicated on the scale on the left. x, Per capita emissions of carbon dioxide, as indicated on the scale on the right. Based on Energy Information Administration 2008b.

Fort Worth, and 6 percent for Houston (Federal Highway Administration 1990–2005). Over the last 20 years these cities have had very high rates of growth, much of it in the suburbs. In recent years, both Houston and Dallas have developed light rail systems and encouraged higher density development in the central city. This, together with higher gasoline prices and high-occupancy vehicle lanes, may have slowed the rate of increase in VMT.

Methane emissions are hard to quantify. Although small, they are an important component of greenhouse gas emissions, because a ton of methane is equivalent to almost 2.5 tons of CO_2. According to the Texas Natural Resource Conservation Commission report published in 2002, methane accounted for 7.4 percent of carbon-equivalent greenhouse gas emissions in 1999, slightly less than the 8.3 percent in 1995 and 1990. The largest contributors are landfills, enteric fermentation, and natural gas extraction and transmission. During the study period, there was a significant decline in emissions from natural gas production and a slight decrease as a result of programs to capture methane at landfills.

Nitrous oxide is another small but significant contributor to global warming. It declined from 4.5 percent of total carbon-equivalent emissions in 1990 to 2.7 percent in 1999 (TNRCC 2002), largely as a result of efforts to improve air quality.

POLICY OPTIONS FOR REDUCING GREENHOUSE GAS EMISSIONS
Reducing Energy Consumption and Air Pollution

Reducing greenhouse gas emissions typically involves reducing energy consumption, often by making equipment and vehicles more fuel-efficient or by improving building standards. Many such programs were initiated in the 1970s following the 1973 Arab oil embargo. In 1975 Congress passed the Energy Policy and Conservation Act, which included the creation of a national infrastructure of state energy offices and set broad national goals. Some funding was provided, but implementation was left to the states, which were required to document energy savings (NASEO 1991). President Carter was also a proponent of energy conservation, seeing our dependence on foreign oil as a national security issue. With lower oil prices in the 1980s, many of these programs declined in importance.

At the time many of these programs were initiated, global warming was not considered to be an issue. The 1990 amendments to the Clean Air Act were primarily aimed at reducing pollution, particularly that causing acid rain. New stationary sources, such as power plants, came under strict new regulations, and vehicle emission standards were tightened. In addition, states were required to improve air quality, and they lost highway funding if cities did not meet certain air quality standards. This resulted in efforts to increase carpooling and encourage the use of public transportation. A recent decision by the U.S. Supreme Court has also provided the EPA with the authority necessary to regulate greenhouse gases under the Clean Air Act.

Individual states have the authority to promote energy efficiency through utility regulation, building codes, mass transit, urban planning, and state taxes and rebates. States also have the ability to regulate vehicle emissions, encourage sequestration of CO_2, and promote fuel switching. Through statewide education programs, states can inform the public about the threat of global warming and the part that the individual can play in dealing with the problem. In addition, individual state governments can set an example through their own purchasing policies (e.g., buying recycled and recyclable materials and ensuring that state car fleets use the most fuel-efficient vehicles). In 1989 the EPA awarded a grant to the National Association of State Energy Officials to evaluate some of these programs. The most successful programs were found to be cost-effective, frequently with a pay back period considerably less than the life of the project. Under these conditions, the programs can be justified on economic terms alone, and the savings of CO_2 emissions are realized at no cost (NASEO 1991).

Motor vehicles produce a variety of greenhouse gases. The major component is CO_2, but several other components of vehicular emissions are produced in significant quantities. The most important of these, for the purposes of the current discussion, are nitrogen oxides and volatile organics, which combine to form ozone. Ozone is a major component of smog and also contributes to global warming by acting as an infrared absorber. Another component, carbon monoxide, prolongs the life of methane in the atmosphere and is eventually converted to CO_2.

Currently, manufacturers of new passenger cars sold in the United States are subject to CAFE standards, which require that vehicles meet an average fuel economy standard of 27.5 mpg. The first edition of this book (in 1995) mentioned a proposal in Congress to increase this level to 40 mpg by the year 2001. Twelve years later, in 2007, Congress finally acted to raise the standard to 35 mpg by 2020. Economic analyses have shown that there are less expensive ways to achieve energy conservation and reduce greenhouse gas emissions than CAFE standards. One study estimated that a gasoline tax of 27.5 cents imposed over five years would achieve the same petroleum and CO_2 savings through 2003 at less than one quarter the cost (Charles River Associates 1991). It is likely that when higher oil prices return, especially if they exceed $100 per barrel, gasoline price increases will have a more dramatic effect than increasing CAFE standards to 35 mpg, making fuel efficiency a significant factor in consumers' choice of vehicles.

Alternative fuels such as ethanol were initially blended with gasoline in small amounts to reduce air pollution. More recently they are being seen as a means of reducing greenhouse gases from transportation. These fuels vary in their effectiveness in reducing such gases, and that effectiveness is highly dependent on how they are produced. Fuels that reduce emissions include natural gas, electricity generated by natural gas plants or renewable energy (for electric vehicles), and fuels produced from biomass (methanol or ethanol). Use of solar energy and fuel cells can also produce net reductions in greenhouse gas emissions, but their use is dependent on continued technological improvements. Results of the use of alcohol fuels (methanol and ethanol) range from a small improvement to an increase in greenhouse gas emissions, depending on how they are produced. Methanol is generally produced from natural gas and results in a slight reduction in greenhouse-related emissions. Ethanol produced from biomass greatly reduces emissions, but ethanol produced from corn or grain, as is the general practice, may actually result in an increase. Even though it is often claimed that greenhouse gas emissions are reduced by 20 percent, a recent study shows that emissions could increase by as much as twofold when all of the environmental consequences of ethanol production (including deforestation to increase the acreage available for crops) are considered (Searchinger 2008).

Programs in Texas

In 2005 Texas consumed 278 million barrels of motor gasoline, 8.3 percent of the total U.S. consumption. As a result, Texas emitted 113 million short tons of CO_2 (16.2 percent of Texas CO_2 emissions; Table 8.1). Several mechanisms are available for reducing greenhouse gas emissions from vehicles, including reducing the amount of fuel consumed and controlling gaseous emissions. The former includes increasing the fuel efficiency of vehicles, encouraging multiple occupancy of vehicles, and considering a wider range of criteria in transportation planning. Although Texas was one of the first states to develop a ride-share program, it has given it little attention since the mid-1980s, when low oil prices resulted in a reduction in public demand. Other incentives include higher parking fees and lanes reserved for high-occupancy vehicles.

The State Energy Conservation Office is a small agency with a mission to reduce energy consumption. Most of its focus is on education, but there are limited funds for energy conservation projects. Over the past several years, the office has funded solar, wind, and biomass demonstration projects throughout Texas and has cosponsored conferences, workshops, and other educational efforts, yielding a positive impact on Texans' familiarity with renewable energy. Free training was provided to public housing authorities in Houston, Dallas, and San Antonio so that they could learn how energy performance contracting could benefit their complexes. It is hoped that as a result of these workshops, 20 or more of these housing authorities will invest $80–$100 million over the subsequent three years for energy-related capital improvements.

The Energy Conservation Office has also partnered with the Housing Trust Fund Program, of the Texas Department of Housing and Community Affairs, to increase the energy efficiency of new and rehabilitated multi- and single-family housing for low- to moderate-income families. The Housing Trust Fund Program is the only state-authorized project dedicated to the development of affordable housing, providing loans to finance, acquire, rehabilitate, and develop affordable housing for low- and very low-income families. The Energy Conservation Office provided funds to ensure that energy-efficient design and appliances were incorporated into new housing construction, requiring projects to exceed the Model Energy Code of 1992 and 1995.

The Lone Star revolving loan program has saved taxpayers more than $200 million through energy-efficiency projects for state agencies, institutions of higher education, school districts, county hospitals, and local governments. Borrowers repay loans through cost savings generated by the projects. Loan Star–funded projects have prevented the release of 7,130 tons of nitrogen oxides, 2.1 million tons of CO_2, and 4,832 tons of sulfur dioxide.

The agency also administers a program that sets minimum requirements for energy efficiency standards for new state buildings or those undergoing major renovation. The standard is based on the American Society of Heating, Refrigeration, and Air Conditioning Engineers Standards 90. The goal is to construct, operate, and maintain buildings in such a way as to reduce energy consumption without compromising the function, productivity, or comfort of the occupants. The State Energy Conservation Office also provides energy code training throughout the state as part of a larger program to improve the energy efficiency of new buildings, including single-family homes. Legislation passed by the Texas Legislature in 2001 adopts the energy-efficiency chapter of the International Code Council's International Residential Code as the standard municipal residential building code for the state. Municipalities may adopt local amendments, but these may not result in less stringent energy-efficiency requirements for any areas designated as nonattainment for air quality standards as designated by the EPA. Local amendments must be reviewed by the Energy Systems Laboratory of the Texas A&M University System, designated by the Legislature as the agency responsible for energy code implementation. In the first year, the Texas Building Energy Performance Standards were projected to save nearly 1.75 trillion Btu from residential buildings alone, with approximately $1.1 billion in annual energy cost savings by 2008. Greater energy efficiency will reduce emissions from power plants fired by coal and natural gas. Public Citizen estimates that by 2008, the annual reduction in nitrogen oxide emissions is expected to exceed 21,000 tons, which will help Texas nonattainment areas meet Clean Air Act requirements (Office of Energy Efficiency and Renewable Energy 2002).

Because the chemical sector relies extensively on natural gas as both feedstock and fuel, energy costs account for a large percentage of manufacturing costs, putting this sector at a disadvantage in a global market. Texas Industries of the Future is a partnership strategy in which the Industrial Technologies Program of the U.S. Department of Energy and the Texas Energy Conservation Office provide outreach, technical assistance, and training for Texas industrial energy consumers. It conducts conferences, workshops, and forums, providing training and outreach to engineers and consultants in process industries on a variety of industrial energy-efficiency topics. The goal of this program is to work with the chemical and refining sector to achieve an overall reduction in energy intensity of 15 percent between 2002 and 2010 (State Energy Conservation Office 2008).

Initiatives to Reduce Greenhouse Gas Emissions at the State Level

Many states, notably California, have taken the lead in developing strategies to tackle global warming. The Pew Center on Global Climate Change gives details on its website of policies adopted at the state level. Actions by individual states include one or more of the following.

Monitoring Greenhouse Gas Emissions

On May 8, 2007, more than 30 states plus two Canadian provinces signed on as charter members of the Climate Registry, a collaboration aimed at developing a common system for entities to report greenhouse gas emissions. The registry will serve as a tool to measure, track, verify, and publicly report greenhouse gas emissions consistently and transparently between states. Voluntary, market-based, and regulatory greenhouse gas emissions reporting programs are all supported under the registry. As of August 2008, there are 39 states, nine Canadian provinces, and six states in Mexico included in the registry. Texas is not a participant.

Establishing Emission Reduction Targets

Each state that is aiming to reduce its greenhouse gas emissions has established its own set of guidelines (Table 8.2). In addition, states have formed regional compacts in order to initiate a coordinated effort to address emission reductions.

The Memorandum of Understanding for the Northeast Regional Greenhouse Gas Initiative (RGGI) was the first multistate greenhouse gas emissions cap-and-trade program in the United States. As members of RGGI, states agree to a regional cap-and-trade program covering power plant CO_2 emissions. RGGI aims to cap these emissions approximately at current levels between 2009 and 2015 and then reduce that level by 10 percent by 2019. Currently there are 10 member states.

Originally formed in February 2007 by five western states, including California, the Western Climate Initiative now consists of seven states and four Canadian provinces. It has set a regional goal of reducing heat-trapping pollution to 15 percent below 2005 levels by 2020. An additional 13 jurisdictions, including six Mexican border states, participate as observers.

TABLE 8.2. Emission reduction targets enacted by states

STATE	BASE YEAR	EMISSION REDUCTION
Arizona	2000	2000 levels by 2020; 50% below by 2040
California	1990	1990 levels by 2020; 80% below by 2050
Colorado	2005	20% below by 2020; 80% below by 2050
Connecticut[a]	1990	1990 levels by 2010; 10% below by 2020
	2001	75–85% long term
Florida	1990	1990 levels by 2025; 80% below by 2050
Hawaii	1990	1990 levels by 2020
Illinois	1990	1990 levels by 2020; 60% below by 2050
Maine	1990	1990 levels by 2010; 10% below by 2020
	2003	75–80% long term
Minnesota	2005	15% below by 2015; 30% below by 2025; 80% below by 2050
New Jersey	1990	1990 levels by 2020
	2006	80% below by 2050
New Mexico	2000	2000 levels by 2012; 10% below by 2020; 75% below by 2050
New York	1990	5% below by 2010; 10% below by 2020
Oregon	1990	10% below by 2020; 75% below by 2050
Rhode Island	1990	1990 levels by 2010; 10% below by 2020
Utah	2005	2005 levels by 2020
Virginia	2007	30% below by 2025
Washington	1990	1990 levels by 2020 25% below by 2035; 50% below by 2050

[a] Massachusetts, New Hampshire, and Vermont have the same requirements

Source: Pew 2008

In November 2007, six states and one Canadian province established the Midwestern Regional Greenhouse Gas Reduction Accord. The members agree to establish regional greenhouse gas reduction targets, including a long-term target of 60–80 percent below current emission levels, and to develop a multisector cap-and-trade system to help meet the targets. Participants will also establish a tracking system for greenhouse gas emission reductions and implement other policies, such as low-carbon fuel standards, to aid in reducing emissions. The first midwestern regional agreement among U.S. states to reduce greenhouse gas emissions collectively, it includes a provision for member states and provinces to establish a cap-and-trade system to aid in meeting their regional greenhouse gas emissions target of 15 percent below 2005 levels by 2020. Draft recommendations on the cap-and-trade program were released in January 2009, with a final version submitted to the advisory group in October.

Montana, a member of the accord, has also adopted an emissions standard for new electricity plants, prohibiting the state public utility commission from approving new electric generating units primarily fueled by coal, unless a minimum of 50 percent of the CO_2 produced by the facility is captured and sequestered. Several other states also have specific requirements for power plants. California caps emissions from electricity retailers, and Washington and Oregon require new plants to offset 20 and 17 percent, respectively, of anticipated CO_2 emissions.

Establishing Renewable Energy Targets

More than half of the states have developed renewable energy standards, requiring a certain percentage of electrical generating capacity to come from renewable sources (Table 8.3). In addition, the Western Governors' Association (representing 19 western states, including Texas) unanimously resolved to examine the feasibility and actions required to reach a goal of 30,000 megawatts of clean energy by 2015 and a 20 percent improvement in energy efficiency by 2020.

In 1999, electricity industry restructuring legislation was enacted in Texas. The first renewable portfolio standard (RPS) under that act required electricity providers (competitive retailers, municipal electric utilities, and electric cooperatives) collectively to generate 2,000 megawatts of additional renewable energy by 2009. Each provider was required to obtain new renewable energy capacity based on its market share of energy sales, multiplied by the renewable capacity goal. For example, a competitive retailer with 10 percent of the Texas retail electricity sales in 2009 would be required to obtain 200 megawatts of renewable energy capacity.

TABLE 8.3. Examples of renewable energy targets enacted by states

STATE	RENEWABLE ENERGY TARGETS (% ELECTRICAL GENERATING CAPACITY FROM RENEWABLE SOURCES)
California	20% by 2010
Delaware	20% by 2019
Illinois	25% by 2025
Minnesota	25% by 2025
Missouri	15% by 2021
Montana	15% by 2015
New York	25% by 2013
North Carolina	12.5% by 2021
Ohio	25% by 2025
Oregon	25% by 2025
Vermont	25% by 2025
Washington	15% by 2020

Source: Pew 2008

The Texas RPS was so successful that its 10-year goal was met in just over 6 years. Wind power development in Texas has more than quadrupled since the RPS was established. As a result of competitive pricing, available federal tax incentives, and the state's immense wind resources, wind power is expected to remain competitive with coal- and gas-fired plants. In 2006, Texas brought online the second and third phases of the mammoth Horse Hollow Wind Energy Center, bringing the total project capacity up to 736 megawatts and making it the largest wind farm in the world. That same year Texas surpassed California as the country's largest wind energy producer (Energy Information Administration 2008a), and in 2007 it increased its wind power capacity by 59 percent, more than any other state. As of December 2009, Texas had 9,410 megawatts of wind power capacity, the largest of any state and approximately one quarter of the nation's total capacity (American Wind Energy Association 2009).

In 2005, the Texas Legislature passed a major extension and expansion of the RPS legislation, increasing the state's RPS goal to 5,880 megawatts by 2015; of that, 500 megawatts must come from nonwind resources. This provision indirectly promotes solar power and biomass in Texas and provides farmers and ranchers with new revenue sources from the use of crops and animal waste to produce energy. The legislation also set a goal of 10,000 megawatts in renewable energy capacity by 2025. The bill included a provision to increase transmission capacity, to get clean energy (especially wind power) from remote areas to the cities (Pew 2008). The Public Utility Commission has since approved a $4.93 billion plan to build transmission lines to carry up to 18,456 megawatts of wind power from West Texas and the Texas Panhandle to metropolitan areas of the state. A third of this capacity already exists and cannot always be utilized because of limitations in the lines (U.S. Department of Energy 2008).

Developing Energy Efficiency Standards and Providing Financial Incentives to Install New Systems

In the face of federal inaction on this issue, California has been a leader in the fight to reduce greenhouse gas emissions. Among other measures, it provides financial incentives for the installation of solar panels for both power generation and hot-water production, and it requires existing and new state buildings and parking facilities to install solar energy systems, where feasible. The California Energy Commission is required to adopt energy efficiency standards for general-purpose lights, with the expectation that the use of incandescent light bulbs in the state will be phased out. A recent $2 increase in vehicle registration fees will fund the new Alternative and Renewable Fuel and Vehicle Technology Program, which will provide financial assistance to develop and deploy low carbon fuels and vehicles.

In 1999, Texas required electric utilities to offset 10 percent of load growth through end-use energy efficiency. In 2007, the Legislature doubled the standard to 20 percent of load growth. Higher targets of 30 percent and 50 percent are being investigated as potential options for the future.

Developing Higher Emissions Standards for Vehicles

California is uniquely qualified to develop improved vehicle emission standards as a result of a provision in the 1971 Federal Clean Air Act. At the time the act was passed, California already had more stringent requirements for improving air quality. The state was given special authority under the act to set its own vehicle emissions standards that go beyond federal standards, though it must first obtain a waiver from the U.S. Environmental Protection Agency. Other states may choose to comply with either the federal or California standard.

On November 8, 2007, the state of California sued the EPA for its failure to act on the state's waiver request, made by the California Air Resources Board in December 2005. California's proposed greenhouse gas emissions standards for motor vehicles would be gradually phased in starting in 2009, and by 2016 the state would require reductions of tailpipe greenhouse gas emissions from new motor vehicles of approximately 30 percent. On December 19, 2007, EPA administrator Stephen Johnson denied California's request for a waiver, arguing that recently enacted federal energy legislation established national vehicle efficiency standards of 35 mpg by 2020, and that this unified standard was preferable to a state-by-state approach. On January 2, 2008, California filed a lawsuit challenging the EPA's decision. Fifteen other states intend to adopt the California standard and are parties to the lawsuit (Roosevelt 2008).

Seeking to resolve the dispute and increase the fuel efficiency of American vehicles, in May 2009 President Obama proposed a new national fuel economy program that adopts uniform federal standards to regulate both fuel economy and greenhouse gas emissions while preserving the legal authority of the Department of Transportation, the EPA, and California. Final rules, issued April 1, 2010, by the EPA, the Department of Transportation, and the National Highway Traffic Safety Administration require automakers, starting with 2012 model year vehicles, to improve fleetwide fuel economy and reduce fleetwide greenhouse gas emissions by approximately 5 percent every year. The EPA standards require manufacturers, by the 2016 model year, to achieve a combined average vehicle emission level of 250 grams of CO_2 per mile, the equivalent of 35.5 mpg, up from the current average for all vehicles of 25 mpg (EPA 2010).

U.S. Mayors Climate Protection Agreement

Recognizing that climate change is an urgent threat to communities across the world and that the United States was not likely to address the problem at the national level, Mayor Greg Nickels of Seattle launched the U.S. Mayors Climate Protection Agreement. Seattle was well placed to lead this effort. It already had policies in place that would help to reduce greenhouse gas emissions, including a goal of net-zero emissions at the city-owned utility, to be achieved through a combination of renewable energy sources and carbon offsets. The city has invested heavily in mass transit and has educational programs in place to encourage energy conservation in heating and lighting, as well as encouraging people to cycle and walk. In addition, the Seattle Climate Partnership was developed to encourage the private sector to reduce its emissions.

The U.S. Mayors Climate Protection Agreement, announced the day the Kyoto Protocol became law in 141 countries, has since been signed by more than 1,000 mayors, representing more than a quarter of the U.S. population. More than 20 Texas mayors have signed the agreement, including those in Dallas, Fort Worth, San Antonio, and Austin. Under the agreement, participating cities commit to take the following three actions:

- Strive to meet or beat the Kyoto Protocol targets in their own communities, through actions ranging from anti-sprawl land-use policies to urban forest restoration projects to public information campaigns.
- Urge their state government and the federal government to enact policies and programs to meet or beat the greenhouse

gas emission reduction target, suggested for the United States
in the Kyoto Protocol, of 7 percent below 1990 levels by 2012.

• Urge the U.S. Congress to pass the bipartisan greenhouse gas
reduction legislation, which would establish a national emission trading system.

On November 1–2, 2007, 100 mayors attended the 2007 Mayors
Climate Protection Summit in Seattle, Washington. At the summit, the
U.S. Conference of Mayors and the Clinton Foundation's climate initiative
announced a new partnership. All 1,100 cities in the Conference of Mayors
will have the opportunity to purchase energy-efficient and clean-energy
products and technologies through the existing purchasing consortium of
the Clinton Climate Initiative. Mayors at the conference also urged Congress
to complete energy efficiency legislation by the end of the year. The city
of Houston is also participating in the Energy Efficiency Building Retrofit
Program, a partnership between the Clinton initiative, four multinational
energy service companies, five global banks, and 16 major cities around
the world to reduce energy use significantly in municipal buildings. Under
the program, participating cities will retrofit their municipal buildings with
more efficient heating, cooling, and lighting systems, reflective roofs, and
other efficiency measures and products. These upgrades are expected to
reduce energy use in the buildings by 20-50 percent. The five participating
banks (ABN AMRO, Citi, Deutsche Bank, JPMorgan Chase, and UBS) have
agreed to contribute $1 billion each to help finance the project, and cities
will repay the bank loans with their energy cost savings (Pew 2008).

The City of Austin prides itself on being a leader in reducing fossil fuel
consumption, and its green building program has been a model for the
nation. The program seeks to educate both builders and the public on energy
conservation and sustainable building techniques, as well as rating new
homes according to a point system that results in the assignment of up to five
stars. Its Green Choice program for electricity generation from renewable
sources has also received national recognition. In February 2007 Mayor Will
Wynn unveiled an aggressive plan to increase the community's commitment
to reductions in greenhouse gas emissions. It has five components: municipal
operations, electric utility planning, homes and buildings, community planning, and carbon-neutral lifestyle.

Municipal Operations

This initiative strives to make all of the city's internal municipal functions
100 percent carbon-neutral by 2020. All city-owned and -operated buildings

and other facilities will be exclusively powered by 100 percent renewable energy by 2012. Only wind, solar, and biomass power will be used. All city vehicles will be carbon-neutral by 2020. The intent is to have the entire fleet, even heavy equipment, powered by electricity or nonpetroleum fuels, if technically possible. Within every individual department, the city will achieve the maximum possible reductions of greenhouse gas emissions and energy consumption. The city will educate, motivate, and support its more than 10,000 employees to help them reduce their personal carbon footprints.

The Utility Plan

Austin Energy has set a goal of becoming the top utility in the nation for greenhouse gas reductions. The city-owned electric utility is already a national leader in environmental initiatives, including its Green Building Program, Green Choice Renewable Energy Program, and Plug-In Partners, a national initiative promoting advanced hybrid vehicles to automakers and fleet purchasers. Through new improvements in efficiency and conservation, Austin Energy will reduce energy use from current levels by a total of 700 megawatts (the equivalent of a whole power plant) by 2020. In addition, Austin Energy has committed to meet 30 percent of all energy needs through renewable energy by 2020.

Homes and Buildings

Because buildings are responsible for 70 percent of emissions, new buildings will be required to meet strong energy-efficiency measures amended into (and enforced through) the city's building code. These measures expand on the energy-efficiency components of the city's successful Green Building Program but take them much farther. Four specific initiatives include a goal to make all new single-family homes net-zero energy capable by 2015. That means that home builders will be required to increase energy efficiency by 65 percent. This ambitious goal would be roughly equivalent to achieving the highest possible energy score for a top five-star rating within the existing Green Building Program. For all other new construction, the goal is to increase energy efficiency by 75 percent by 2015. In addition, when a house or other building is resold, it will be required to have an energy-efficiency inspection and to meet minimal standards. The buyer will be responsible for adding basic improvements, if necessary, and these can be financed through Austin Energy's zero-interest loan program or rolled into a 30-year mortgage.

Community Plan

Although less well defined, one Austin goal is to involve the whole community through a broad public-information campaign that will lead to innovative methods for inspiring behavioral change. Two major areas of concern are transportation and land-use planning. The transportation initiative includes a two-pronged approach to get people out of their cars through investments in mass transit, while also putting drivers into lower-emission vehicles (the goal of Austin Energy's Plug-In Partners program). The land-use effort includes encouraging urban infill to promote density rather than suburban sprawl, which in turn supports greater use of public transit and reduced car-trip miles.

Going Carbon Neutral

The city will also develop and implement a program to motivate and help everyone in Austin (individuals, businesses large and small, organizations, even visitors) voluntarily take their net greenhouse gas emissions down to zero. Those who aspire to a carbon-neutral lifestyle generally improve everything they can by buying energy-efficient appliances and cars and then purchasing compensating offset credits to zero-out their remaining greenhouse gas emissions. An online tool will help households and small businesses calculate their total greenhouse gas emissions, based on Austin data (City of Austin 2007).

POLICY OPTIONS FOR TEXAS

State action, however crucial, is not a substitute for federal action. As the report *Reducing the Rate of Global Warming: The States' Role* concludes: "States can initiate models of effective programs, but a strong federal presence is required to ensure that all states implement effective policies" (Machado and Piltz 1988). In addition, there are certain policy options that states cannot take unilaterally and remain economically competitive; these policies are best pursued by the federal government. For example, a charge based on the carbon content of specific fossil fuels would most likely be passed on to the consumer in the cost of finished products and would place an unfair burden on any one state instituting such a carbon tax on its own. (Indeed, such a charge may have to be part of an international system.) Similarly, standards regarding fuel efficiency of vehicles or the energy efficiency of household appliances would be most easily applied at the point of manufacture by the federal government.

As the previous section demonstrates, there is a whole range of energy-saving measures that are cost-effective, even if the only benefit considered is the avoided energy cost. In many cases the projects pay for themselves long before the effective life of the program or equipment expires, and the additional benefit of reducing CO_2 emissions is literally without cost. Often all that is needed to implement such programs is a pool of money for capital expenditures or low-interest loans. Texas has a number of state-financed revolving funds, and these could be used as a model for similar energy-conservation programs.

So far, Texas has done very little to address the problem of global warming. In fact, the official policy appears to be to wait and see what the federal government does. Nevertheless, several cities, particularly Austin, have taken a leading role and will provide models once action is mandated either by the state or federal government. Given that more than half of U.S. states have taken some action, it would be preferable for Texas to take the initiative and not wait for regulations to be imposed on it. With oil prices exceeding $100 per barrel likely to return in the near future, most of the following solutions would be cost-effective within a very short time frame.

- Monitor greenhouse gas emissions: Join more than 30 states and four Canadian provinces as members of the Climate Registry, a collaboration aimed at developing a common system for reporting greenhouse gas emissions.
- Establish emission reduction targets: Such targets could be developed specifically for Texas, or Texas could join the Western Climate Initiative, which has set a regional goal of reducing heat-trapping pollution to 15 percent below 2005 levels by 2020. By joining such an alliance now, Texas would be able to participate in the development of regional solutions such as a cap-and-trade program.
- Increase renewable energy targets: Because of the great potential for wind power, Texas was one of the first states to set goals for renewable energy generation. The current target of 5,880 megawatts by 2015 represents only about 5–7 percent of total generating capacity and has already been exceeded. Increasing the target to 20 percent would be an entirely achievable goal. New and existing state buildings could be required to install solar energy systems.
- Develop energy efficiency standards: As a result of legislation passed in 2007, utility companies are required to meet 20 percent of new demand from energy efficiency. In addition, the Public Utility Commission is required to study the feasibility of increasing this to 50 percent of new energy demand within 10 years. More could be done. For instance, California is developing standards for general-purpose lights.

- Provide financial incentives: The state could provide financial incentives for the installation of energy efficient equipment, such as air conditioners. This could be used to encourage customers to buy units of greater efficiency or to replace existing units. Financial assistance for the purchase of solar panels for the generation of power or hot water would also encourage their installation.
- Improve building codes: Increase the energy efficiency requirements in the standard municipal residential building code.
- Replace roads with rail: Instead of promoting roadway construction for the transportation of NAFTA traffic, develop a more extensive rail system, which would move freight with less than half the fuel used by trucks.
- Develop standards for state agency purchases: Require state agencies to purchase equipment that meets certain minimum efficiency standards and cars that meet certain fuel efficiency standards, including a certain percentage of hybrid and plug-in hybrid vehicles (when available).

With increasing energy costs and fewer resources, at both the personal and government level, saving energy is not only wise environmental policy but also good economics. Reducing input costs for the agricultural and industrial sectors of the state will improve the competitiveness of Texas products in the marketplace. Our competitors in Japan and Europe use less than half the energy we use for manufacturing processes. One of the factors contributing to the high energy intensity of this society is the array of considerable subsidies associated with energy use. Among the many examples of this are government expenditures for defense to protect oil imports, infrastructure maintenance associated with transportation, direct subsidies to oil companies, and the failure to account for the health and environmental costs of air pollution.

As the pressures increase to improve energy efficiency and reduce greenhouse gas emissions, Texas has an opportunity to evaluate the role that it can play in the debate. By taking an active part, Texas can help to shape policy options with its own interests in mind. The Texas economy is heavily dependent on the energy sector. As a major exporter of natural gas, Texas would be the beneficiary of policies aimed at substituting natural gas for other fuels such as coal, which produces twice as much CO_2 for the same amount of energy generated. In the longer term, Texas could position itself to take advantage of opportunities to develop and manufacture new energy-efficient technologies. The potential for wind and solar power in Texas is almost unlimited. The development of these technologies would create well-paying jobs in an industry with great export potential, and encouraging the use of renewable energy for electricity generation would, in the long run, make Texas products more competitive.

The United States, which produces more than 20 percent of worldwide anthropogenic CO_2 emissions, has an obligation to develop policies to limit its reliance on fossil fuels. Reducing CO_2 emissions through energy conservation is a realistic goal. Many states have demonstrated the economic feasibility of such measures, and Texas could take a more aggressive approach to the problem. Until such time as national and international policies are implemented, however, emphasis should be placed on cost-effectiveness and overall economic benefits.

REFERENCES

American Wind Energy Association, 2008. Wind: Powering a Cleaner, Stronger America. *U.S. Wind Energy Projects*. www.awea.org, accessed March 2010.

California Energy Commission, 2006. Inventory of California Greenhouse Gas Emissions and Sinks, 1990 to 2004: Final Staff Report. Publication CEC-600-2006-013-SF, Figures 9 and 11. California Energy Commission, Sacramento.

Charles River Associates, 1991. Policy Alternatives for Reducing Petroleum Use and Greenhouse Gas Emissions. Final Report.

City of Austin, 2007. *Austin Climate Protection Plan*. www.ci.austin.tx.us/council/downloads/mw_acpp_points.pdf.

Energy Information Administration, 2008a. State Energy Profiles: Texas. U.S. Department of Energy, Washington, D.C. http://tonto.eia.doe.gov/state/, accessed June 2008.

Energy Information Administration, 2008b. Tables 7–12. Energy Consumption Estimates by Source and Sector, Selected Years, 1960–2005: Texas. www.eia.doe.gov/emeu/states/sep_use/total/, accessed June 2008.

EPA, 2007. The U.S. Inventory of Greenhouse Gas Emissions and Sinks: Fast Facts. EPA 430-F-07-004. U.S. Environmental Protection Agency, Washington, D.C.

EPA, 2010. DOT, EPA Set Aggressive National Standards for Fuel Economy and First Ever Greenhouse Gas Emission Levels For Passenger Cars and Light Trucks. http://yosemite.epa.gov/opa/admpress.nsf/2010%20Press%20Releases!OpenView, accessed April 1, 2010.

Federal Highway Administration, 1990–2005. Highway Statistics 1990, 1992, 1995, 1997, 2000, and 2005. www.fhwa.dot.gov/policy/ohim/hso5/roadway_extent.htm, accessed June 2008.

Finn, P., 2004. Russia to Adopt Kyoto Protocol. *Washington Post*, September 30, 2004.

Graedel, T. E., and P. J. Crutzen, 1989. The Changing Atmosphere. *Scientific American* 261(3):58–68.

IPCC, 2007a: Summary for Policymakers. In: *Climate Change 2007: The Physical Science Basis.* Contribution of Working Group 1 to the Fourth Assessment Report of the Intergovernmental Panel on Climate Change. Cambridge University Press, Cambridge, U.K., and New York.

IPCC, 2007b: Summary for Policymakers. In: *Climate Change 2007: Impacts, Adaptation, and Vulnerability.* Contribution of Working Group 2 to the Fourth Assessment Report of the Intergovernmental Panel on Climate Change. Cambridge University Press, Cambridge, U.K., and New York.

IPCC, 2007c: Summary for Policymakers. In: *Climate Change 2007: Mitigation of Climate Change.* Contribution of Working Group 3 to the Fourth Assessment Report of the Intergovernmental Panel on Climate Change. Cambridge University Press, Cambridge, U.K., and New York.

Machado, S., and R. Piltz, 1988. *Reducing the Rate of Global Warming: The States' Role.* Renew America, Washington, D.C.

NASEO, 1991. *Energy Efficiency and Renewable Energy: Economical CO_2 Mitigation Strategies.* Report on the Pilot Phase, EPA/NASEO Joint Project. National Association of State Energy Officials, Washington, D.C.

Office of Energy Efficiency and Renewable Energy, 2002. Texas Building Code and Energy Performance Standards. U.S. Department of Energy, Washington, D.C. www.energycodes.gov.

Pew, 2008. U.S. States and Regions. Pew Center on Global Climate Change. www.pewclimate.org/states-regions, accessed June 2008.

Roosevelt, M., 2008. Lawsuit Targets EPA's Refusal; California and 15 Other States Seek to Overturn Agency's Ruling on Tailpipe Emissions. *L.A. Times,* January 3, 2008.

Searchinger, T., R. Heimlich, R. A. Houghton, F. Dong, A. Elobeid, J. Fabiosa, S. Tokgoz, D. Hayes, T. H. Yu, 2008. Use of U.S. Croplands for Biofuels Increases Greenhouse Gases through Emissions from Land-Use Change. *Science* 319(5867):1238

State Energy Conservation Office, 2008. Office of the Texas Comptroller, Austin. www.seco.cpa.state.tx.us/energy-efficiency/, accessed June 2008.

TNRCC, 2002. A Report to the Commission Concerning Greenhouse Gases in Texas. Texas Natural Resource Conservation Commission, Austin.

U.S. Department of Energy, 2008. Texas to Spend $4.93 Billion on Transmission Lines for Wind Power, July 23, 2008. http://apps1.eere.energy.gov/news/news_detail.cfm/news_id=11886, accessed March 2010.

U.S. Mayors Climate Protection Agreement, 2010. www.seattle.gov/mayor/climate, accessed March 2010.

Vidal, J., J. Jowit, and D. Adams, 2007. Hard-won Climate Deal Gets Lukewarm Reception. *Guardian Weekly,* December 21, 2007.

Vidal, J., A. Stratton, and S. Goldenberg, 2009. Low Targets, Goals Dropped: Copenhagen Ends in Failure. www.guardian.co.uk/environment/2009/dec/18/copenhagen-deal, accessed March 2010.

CHAPTER 9

Economy
Jared Hazleton

Earlier chapters have shown that elevated levels of greenhouse gases in the atmosphere and the resulting impacts on the earth's climate could have significant environmental impacts. This chapter focuses on the economic impacts of climate change. It begins by reviewing the latest Texas demographic and economic trends and forecasts, followed by a discussion of the likely economic impacts on the state of projected climate change. The concluding sections discuss the economics of climate change, the types of public policy that can be used to address it, and the economic impact on the United States and Texas of national climate change policy.

TEXAS POPULATION TRENDS

The 2000 Census reported the population of Texas at 20.9 million. The state's January 2009 population was estimated at 24.5 million (Texas State Data Center 2010). Of the four alternative scenarios for population projections presented here (Table 9.1), the 0.5 scenario is recommended as the most appropriate for purposes of long-term planning. This scenario assumes an annual rate of growth approximately 1.5% slower than that experienced in the 1990s; nonetheless, it projects that Texas will add 15.9 million persons (a 71.5% increase) between 2000 and 2040, a numerical increase greater than the state's total population in 1980. By way of comparison, the U.S. Census projects only a 50% increase in the U.S. population between 2000 and 2050.

To understand the potential impacts of climate change on the Texas economy, one must anticipate not just how many Texans there will be but also how they will be distributed across the state. Under the 0.5 scenario, growth is evident across a majority of Texas counties, with 211 of the state's 254 counties showing population increases from 2000 to 2040.[1] Three areas of the state—the four major urban centers, the counties along the Texas Gulf Coast, and the counties bordering on Mexico—are especially important to an understanding of the economic impacts of climate change.

The transportation sector is responsible for a sizable portion of the greenhouse gas emissions in Texas. The number of Texans residing in urban areas is a significant indicator of the potential for these emissions. The

TABLE 9.1. Texas population in 2000 and projected to 2040, under alternative demographic assumptions

YEAR	POPULATION (MILLIONS)				% CHANGE			
	0.0 SCENARIO[a]	0.5 SCENARIO[b]	1990–2000 SCENARIO[c]	2000–2004 SCENARIO[d]	0.0 SCENARIO	0.5 SCENARIO	1990–2000 SCENARIO	2000–2004 SCENARIO
2000	20.9	20.9	20.9	20.9				
2010	22.8	24.3	26.1	25.1	9.4	16.7	25.0	20.4
2020	24.3	28.0	32.7	30.3	6.7	15.1	25.6	20.5
2030	25.5	31.8	41.1	36.3	4.6	13.7	25.6	20.1
2040	26.1	35.8	51.7	43.6	2.5	12.3	25.8	20.0
Change 2000–2040	5.2	14.9	30.8	22.7	25.1	71.5	148.0	109.0

[a] The 0.0 scenario assumes no net migration.

[b] The 0.5 scenario assumes net migration equal to one half of the age, sex, and race-ethnicity–specific rates of net migration experienced in the 1990s.

[c] This scenario assumes net migration equal to that experienced in 1990–2000.

[d] This scenario assumes net migration equal to that experienced in 2000–2004.

Source: Texas State Data Center (2009)

four largest metropolitan areas in Texas (Austin–Round Rock, Dallas–Fort Worth–Arlington, Houston–Sugar Land–Baytown, and San Antonio) had a population of 12.8 million in 2000, representing nearly 59% of the state's population. By 2007, their estimated population had risen 19.5%, to about 15.3 million. These urban areas are projected to have a population of 23.7 million by 2040 (0.5 scenario) and would be home to two out of every three Texans.

The 18 counties along the Texas Gulf Coast are estimated to have added nearly 655,000 persons between 2000 and 2007, a 12.6% increase. Under the 0.5 scenario, the coastal population would rise 60%, from 5.2 million in 2000 to a projected 8.3 million in 2040. This would place additional pressures on shoreline development that might be significantly affected by a continued rise in sea level.

The population of the 13 counties along the Texas-Mexico border is estimated to have risen about 18% between 2000 and 2007. Under the 0.5 scenario, the border population would rise from 2.3 million in 2000 to a projected 4.1 million in 2040, a 78% increase. This region is the poorest in Texas and has major health problems not found in other parts of the state. If, as many anticipate, climate change causes the incidence of tropical diseases to move northward, new health issues are likely to arise. This area is also dependent on water from the Conchos and Pecos rivers, which are likely to experience reduced flows due to projected climate change.

The young and the old represent special populations that are especially vulnerable to a warming climate. Historically, Texans have been younger on average than their counterparts across the nation. For example, the 2000 Census revealed a median age for Texas of 32.3 years, 3 years below the national median age of 35.3 years. The median age in Texas, however, is projected to rise to 37.9 years by 2040. The population 65 years of age and older in 2040 is expected to total 7.1–8.2 million, up from 2.2 million in 2000 (Texas State Data Center 2010).

THE TEXAS ECONOMY

Historically, Texas depended on agriculture and minerals (cotton, cattle, oil and gas) to fuel its economy. Later, refining and petrochemical manufacturing became important. These industries were hit hard by the 1980s downturn in the Texas economy, as oil and gas prices plunged and the state's financial and real estate markets collapsed. By 1990, the share of the goods sector in the state's economy had fallen to about 27% and it is expected to remain at that level through 2030 (Table 9.2).

TABLE 9.2. Texas gross state product, projected by sector

	1990		2000		2030	
SECTOR	BILLIONS OF CHAINED $[a]	% OF TOTAL	BILLIONS OF CHAINED $[a]	% OF TOTAL	BILLIONS OF CHAINED $[a]	% OF TOTAL
Agriculture	4.40	1.0	6.47	0.9	6.62	0.3
Mining (oil and gas)	49.27	10.7	45.18	6.2	28.20	0.1
Construction	22.51	4.9	36.88	5.1	80.80	4.1
Manufacturing	47.25	10.2	92.88	12.8	409.88	21.0
Total goods (constant 2000 $)	123.43	26.7	181.52	25.0	525.51	27.0
Wholesale and retail trade	55.68	12.1	102.0	14.0	242.7	12.5
Transportation and utilities	23.94	5.2	53.79	7.4	90.47	4.6
Other services[b]	197.77	42.8	309.35	42.5	931.53	47.9
Government	63.99	13.9	80.59	11.1	157.27	8.1
Total services (constant 2000 $)	338.57	73.3	545.72	75.0	1,422.00	73.0
Total gross product (constant 2000 $)	462.00	100.0	727.23	100.0	1,738.93	100.0
Total gross product (current $)	382.04	727.23	4,178.53		3,319.51	

[a] The measure *chained dollars* is used to express real dollar amounts adjusted over time for inflation. Chained dollars are based on the average weights of goods and services in successive pairs of years. The measure is chained because the second year in each pair, with its weights, becomes the first year of the next pair. Chained dollars do not sum to category totals.

[b] Includes information, financial activities, professional and business services, education and health services, leisure and hospitality, and other services.

Source: Texas Comptroller 2009

Within the goods sector, however, a second transformation has been taking place. At the height of the oil boom in the early 1980s, energy production accounted for nearly a fourth of the gross state product (GSP; Texas Comptroller 2008). In 1990, however, agriculture and oil combined accounted for only 12% of GSP. By 2000, the share of these two industries had fallen by more than 40%, and this trend is projected to continue. In 2030, agriculture and mining are expected to account for only 0.4% of the state's output. This is significant because national studies indicate the major economic impact of a changing climate over the next few decades is a reduction in agricultural output, while the major part of the cost of reducing carbon emissions would fall on energy-related activities. The other sector impacted by climate change is transportation and public utilities. Output in this sector rose strongly between 1990 and 2000, and is projected to more than double over the next two decades (Table 9.2).

ECONOMIC GROWTH AND GREENHOUSE GAS EMISSIONS

Gross state product allows us to track the growth of the state's economy in real terms (Table 9.2). Real-dollar projections take out the influence of inflation by expressing the projected quantity of goods and services produced in constant 2000 dollars. This is important when comparing GSP growth with growth in greenhouse gas emissions or other pollutants measured in physical units. For a given level of technology, there exists a direct relationship between the amount of economic growth and environmental pollution, including greenhouse gas emissions. More people, more economic activity, more travel, and more demand for fossil fuels: all increase the pressure on environmental and natural resources.

As indicated in earlier chapters, the primary contributors to greenhouse gas emissions in Texas are manufacturing, utilities, oil and gas production, and transportation. Those sectors of the economy that contribute the least amount of greenhouse gases are the service industries, other than transportation. It is significant, therefore, that the sectors of the economy in Texas that are growing the fastest contribute the least to greenhouse gas emissions, as well as to other forms of air and water pollution. This is borne out by the trends in population, output, and carbon dioxide (CO_2) from fossil fuels for the United States and Texas from 1990 to 2006, as well as by the U.S. Department of Energy projections for 2020 and 2030 under the "Business as Usual" scenario (Table 9.3). Although Texas exceeds the nation in emissions per capita and in emissions intensity, those numbers are falling in both the United States and Texas and are projected to continue to fall through 2030.

TABLE 9.3. Carbon dioxide emissions, projected to 2030

VARIABLE	1990	2000	2006	BAU[a] 2020	BAU[a] 2030
U.S. CO_2 emissions[b] (million metric tons)	5,069	5,865	5,890	6,384	6,851
Texas CO_2 emissions[b] (million metric tons)	588.3	691.6	641.6[c]	695.4[d]	746.2[d]
U.S. population (millions)	248.7	281.4	298.8	335.8	363.6
Texas population (millions)	17.0	20.9	23.5	28.0	31.8
U.S. emissions per capita	20.4	20.8	19.7	19.6	18.8
Texas emissions per capita	34.6	33.1	27.3	24.8	23.5
U.S. GDP (billions of 2000 $)	7,113	9,817	11,319	15,984	20,219
Texas GSP (billions of 2000 $)	462	727	868	1,286	1,685
U.S. CO_2 intensity tons/million $	713	582	520	399	339
Texas CO_2 intensity tons/million $	1,273	951	739	541	443

[a] BAU, Business-as-usual reference case

[b] From fossil fuels

[c] Estimated (using growth rate from 2004 to 2005) based on released data for 2005

[d] Estimated by taking the proportion of U.S. emissions to Texas emissions in 2006 and applying it to the U.S. projected emissions for 2020 and 2030.

Sources: U.S. CO_2 emissions: EIA 2008a. Texas CO_2 emissions (1990, 2000, 2006): U.S. Environmental Protection Agency, CO_2 Emissions from Fossil Fuel Combustion, http//epa.gov/climatechange/emissions/state_energyCO2inv.html. U.S. population (2020 and 2030): Census Bureau, U.S. Interim Projections by Age, Sex, Race, and Hispanic Origin, 2004. Texas population: U.S. Census Bureau. U.S. GDP (1990, 2000, 2006): U.S. Department of Commerce, Bureau of Economic Analysis. Texas GSP (1990, 2000, 2006, 2020, 2030): Texas Comptroller.

ECONOMIC IMPACTS OF CLIMATE CHANGE ON TEXAS

In recent years, a number of studies analyzing the impact of climate change on a state or region have been published. These studies generally are limited to describing the types of damages that might be experienced in the context of the economic structure and demographic characteristics of a single state or region without employing an integrative model to estimate the impact on the area's overall economy.

The continued rise in global temperature may be expected to have several effects on the growth and development of the Texas economy. The primary industries directly affected by climate are those closely tied to natural resources, such as agriculture and forestry, which are sensitive to both temperature and precipitation changes, and coastal development, which is impacted by the influence of temperature change on sea level and hurricane intensity. The energy industry, transportation infrastructure, and public health are also impacted by climate change.

Agriculture

As discussed in Chapter 6, studies indicate that agriculture is sensitive to climate change, which affects land and water usages as well as crop and livestock production, but the overall economic impact on the agricultural sector is likely to be small (Nordhaus 1994; Nordhaus and Boyer 2000; Jorgenson et al. 2004). The warming impact will vary across regions of the state, with the Texas High Plains being especially vulnerable.

Energy

Due to its large population and energy-intensive economy, Texas leads the nation in direct energy consumption, accounting for more than one tenth of total U.S. energy use. The mix of energy uses in Texas, however, is very different from the nation as a whole. Industry in Texas consumes about 64% of the state's total end-use energy, as opposed to 38% for the nation (EIA 2008b).

The most recent assessment of the impact of climate change on energy production and consumption in the United States concludes that climate warming will mean reductions in total heating requirements and increases in total cooling requirements, which will vary by region and season (Climate Change Science Program 2007). In general, the changes imply increased demands for electricity, which supplies virtually all cooling energy services but only some heating services.

Analysis by the Electric Reliability Council of Texas (ERCOT) indicates that electricity consumption in the state is closely related to temperature (ERCOT 2006).[2] Texas, like most southern states, uses more electricity for space cooling than for heating. Because capacity must be constructed to match summer peak demand for electricity, peak demand for cooling is highly sensitive to temperature increases. An increase in average temperature is likely to necessitate large increases in generating capacity, unless a greater effort is made to promote conservation through rate restructuring and improvements in the energy efficiency of buildings.

From a baseline in 2000, global warming–induced capacity requirements have been projected for Texas for 2030 and 2050, under low-growth and high-growth scenarios (Table 9.4). In both scenarios, generating requirements grow much more slowly than projected economic output. The Texas Comptroller projects that inflation-adjusted GSP in Texas will increase at an annual rate of 3.3% between 2000 and 2030, while the U.S. Energy Information Administration (EIA) is projecting that the nation's real gross domestic product will grow 2% per year (Texas Comptroller 2008; EIA 2008a). The low-growth scenario assumes that electric generation in the state will increase only 1.5% per year between 2000 and 2050. This is quite conservative, given that in 2006 net generating capacity had already jumped a third since 2000 (Texas Comptroller 2008). The high-growth scenario assumes that electric generating requirements will grow 2.1% per year between 2000 and 2050, the rate of growth actually experienced between

1990 and 2006. The increase in peak demand for electricity due to climate change is assumed to be 20%. This is the low end of the estimate (20–30% increase in capacity above the baseline) for the southern region of the United States (Linder et al. 1987; Smith and Tirpak 1990) and slightly higher than a more recent estimate of a 17% increase in peak electricity demand in California (Miller et al. 2008).[3]

TABLE 9.4. Potential impacts of climate change on Texas electric utility capacity

		2030		2050	
CAPACITY	2000	LOW GSP 1.5%/YEAR	HIGH GSP 2.1%/YEAR	LOW GSP 1.5%/YEAR	HIGH GSP 2.1%/YEAR
Baseline capacity (megawatts)[a]	76,000	102,436	141,770	159,999	214,833
Baseline plus 20% for global warming (megawatts)[a]	76,000	122,923	170,124	191,998	257,800
Global warming requirements (megawatts)[b]		20,487	28,354	31,999	42,967
Cumulative cost for additional capacity (billions of $)[c]		21.1	30.0	33.0	45.1

[a] ERCOT 2006

[b] Linder et al. 1987; Smith and Tirpak 1990; Miller et al. 2008

[c] OECD, 2005. Assumes $1,050 per megawatt for the high case and $1,032 per megawatt for the low case (2003 dollars)

Given these assumptions, it is estimated that the cumulative cost of construction of the additional generating capacity induced by climate change between the years 2000 and 2050 will range from $21 billion to $45 billion under the alternative scenarios. These construction costs are in constant dollars and therefore do not include the effects of inflation. Average construction costs per kilowatt of capacity were estimated for both scenarios, using projected costs for new generating facilities in constant 2003 dollars per kilowatt hour (OECD 2005). It is assumed under the high-growth scenario that the current fuel mix will continue unchanged. This gives a weighted cost of $1,050 per generating unit, made up of a composite fuel source consisting of 49% gas, 37% coal, 10% nuclear, and 4% hydroelectric, wind, and other renewables. The low-growth scenario reflects adjustments in the fuel mix as a consequence of tighter regulation of greenhouse gas emissions. It gives a weighted cost of $1,032 per generating unit, comprising 60% natural gas, 20% coal, 15% nuclear, and 5% renewables.

The annual cost of operations for electric utilities is estimated to be about 20% of cumulative capacity cost (Linder et al. 1987). Using this estimate, operating costs for the Texas utility sector in response to global warming could approach $9 billion per year, to cover variable costs, depreciation, and other fixed costs. Generating capacity is required to meet peak (summer) demand. Thus, a warmer summer accompanied by a warmer winter will have the effect of increasing the excess capacity during nonpeak periods. This increase in excess capacity and change in the distribution of seasonal demand during the year adds to the cost of operation.[4]

Increases in energy costs, such as those experienced in 2008, make increases in energy efficiency even more attractive. Price increases spur the development of more advanced technologies, such as more efficient generating and transmission systems. Investment in improved insulation, more efficient air conditioners, alternative construction designs, and other methods of reducing cooling costs may also be stimulated by rising temperatures and increased electricity costs.

Climate warming also will impact energy production and supply if extreme weather events become more intense. In addition, temperature increases reduce overall thermoelectric power generation efficiencies and may impact facility siting decisions. Most of these impacts are thought to be modest, however (Climate Change Science Program 2007).

Indirectly, climate change is likely to affect the investment behavior of some energy institutions, as well as energy technology, investments in R&D, and energy resource and technology choices. Indirect effects include impacts of climate change on other countries in ways that affect U.S. energy conditions through their participation in global and hemispheric energy markets. Climate change concerns could add momentum to efforts aimed at improving U.S. energy security. Given the prominence of the state in energy production, such impacts would likely be significant for Texas energy producers. With the current state of knowledge, however, it is not possible to quantify these indirect impacts (Climate Change Science Program 2007).

Coastal Development

Texas has over 367 miles of coastline and more than 3,300 miles of bay shore, all of which provide a wide range of essential goods and services. These include overlapping and often competing uses, such as recreation and tourism, coastal development, commercial fisheries, aquaculture, biodiversity, marine biotechnology, navigation, and mineral resources.

Although the 18 counties bordering the Gulf of Mexico constitute only 7.2% of the land in the state, they account for nearly a quarter of its population. These areas become more crowded every year. It is projected that the state's coastal population will jump 60% by 2040 (Texas State Data Center 2010). This growth, as well as increased income and wealth, will place rapidly increasing demands on coastal and marine resources.

The Gulf Coast is vital to the state's economy. More than half of the nation's chemical and petroleum production is located on the Texas coast, and the state leads the nation in marine commerce, with 10 deep-draft ports (3 of the top 10 in the nation, based on total cargo tonnage). Each year, cargo with a commercial value of more than $25 billion moves along the 420 miles of the Texas portion of the Intracoastal Waterway. Texas' commercial fishing fleets bring in more than $170 million of fish and shellfish annually. Coastal tourism accounts for more than $7.5 billion a year, representing about one quarter of the money spent annually on travel in the state (Texas General Land Office 2005).

In addition to direct economic benefits, coastal and marine ecosystems, like all ecosystems, have unique properties or processes that directly or indirectly benefit human populations (National Assessment Synthesis Team 2000). Coastal and marine ecosystems are viewed as among the most valuable to society. The dollar value per acre of these systems has been estimated at $9,240 for estuaries, $4,043 for tidal marshes, $2,459 for coral reefs, and $1,640 for coastal oceans (Constanza et al. 1997).[5] Ecosystem valuation is difficult and fraught with uncertainties, but the magnitude of these estimates serves to emphasize their economic importance.

Increasing population and related land use changes have led to major reductions in coastal wetlands (Moulton et al. 1997). The Texas coast may be the most vulnerable region in the nation. Sea-level rise especially impacts Texas because of its low-lying coast, increasing incidence of subsidence, and high rates of erosion due to reduced sedimentation and greater flushing rates. Sea level is projected to rise an additional 19 inches by 2100, with a possible range of 5–37 inches along most of the U.S. coastline (National Assessment Synthesis Team 2000).

Storm Impacts on Infrastructure

As demands on coastal and marine resources increase and as coastal areas become more developed, the vulnerability of human settlements to hurricanes, storm surges, and flooding events also increases (IPCC 2007b). The more property and people that are in the path of a hurricane, the higher the damages and deaths (Pielke and Landsea 1998). As sea levels rise, even if the intensity of storms remains stable, the same hurricane results in greater damages and deaths from storm surges, flooding, and erosion (Pielke and Pielke 1997). Climate models indicate that rising sea surface temperatures may cause significant increases in storm intensity (Emanuel 2005; IPCC 2007b).

Industry, energy, and transportation works are sensitive to weather extremes that exceed their safety margins. Costs of these impacts can be high. For example, power outages in the United States cost the economy $30–$130 billion annually (EPRI 2003; LaCommare and Eto 2004; EEI 2005).

Texas is especially vulnerable because it borders on the Gulf of Mexico. The impacts of Hurricanes Katrina, Rita, and Wilma in 2005 and Ivan in 2004 demonstrated that the Gulf of Mexico offshore oil and natural gas platforms and pipelines, petroleum refineries, and supporting infrastructure can be seriously harmed by major hurricanes, producing national-level impacts and requiring recovery times stretching to months or longer (EEA 2005; Levitan and Associates 2005; RMS 2005). Hurricane Ike devastated the upper Texas coast in the summer of 2008, causing an estimated $24.9 billion in damages (Berg 2009).

THE ECONOMICS OF CLIMATE CHANGE

Economics has an important role to play in helping us understand climate change and in developing effective public policies to address it. Economics can explain why markets have difficulty dealing with the threat of climate change. It can also assist policymakers in balancing the magnitude of the damages from projected climate change against the cost of actions to reduce greenhouse gas emissions.

Climate Change Results from Market Failure

Humans use the atmosphere as a resource to absorb wastes (greenhouse gases) that result from growing crops; raising animals for food and fiber; and heating, cooling, and running machines with fossil fuels. As long as these emissions do not exceed the carrying capacity of the resource, they do not accumulate over time in the atmosphere. Since no one has property rights to the atmosphere, however, the cost of dumping wastes into it is essentially zero. This can lead to overuse, allowing greenhouse gases to build up in the atmosphere to critical levels.

Economics defines an *externality* as occurring when a decision by one party causes costs or benefits to a third party. If participants in an economic transaction do not bear all of the costs or reap all of the benefits of the transaction, too much or too little of the good may be produced and consumed in terms of the overall cost or benefit to society. For example, emitting greenhouse gases in the process of producing electricity can harm people around the world, if it adds to the atmospheric concentration of these gases, causing global temperatures to rise and other climatic change. The social cost of carbon is the present value of additional economic damages now and in the future caused by an additional ton of greenhouse gas emissions.[6]

Because it is the sum of all the greenhouse gases emitted around the earth that matters (i.e., the damage is not related significantly to the location and timing of emissions), the atmosphere may be thought of as a global public good. A public good has two principal characteristics: it is difficult to prevent people from using the good without paying for it (consumption is nonexcludable), and the incremental cost of allowing more users is near zero (consumption is nonrival). As a resource, the atmosphere satisfies both conditions. Market institutions do not function properly when a resource has these characteristics. As a result, the resource may not be efficiently used.

It is customary to look to government to solve the problems posed by the existence of public goods. The solution most commonly proposed for projected climate change is for governments to take action to reduce greenhouse gas emissions sufficiently to restore the atmosphere to a sustainable level of greenhouse gas concentration. The key to effective public policies to address climate change is to internalize the externality caused by greenhouse gas emissions. In practice, this will require that the market price of carbon be increased, which will raise the prices of fossil fuels and the products of fossil fuels. This will transmit the social costs of greenhouse gas emissions to the everyday decisions of billions of people and firms.

Because the causes and consequences of climate change are global, effective mitigation policies require extensive cooperation among countries. This is crucial because, even though the industrialized world today accounts for the largest portion of greenhouse gas emissions, developing countries such as China, India, and Brazil are rapidly joining their ranks and will soon become the largest emitters. In the absence of mitigation policies, it is estimated that two thirds to three quarters of the projected increase in global CO_2 emissions will occur in developing countries (IPCC 2007b).

Even though climate change affects all nations, some may be reluctant to reduce greenhouse gases voluntarily, realizing that they cannot be prevented from enjoying a better climate whether they contribute to it or not (i.e., free-riding). The incentive to free-ride is likely to be much higher in developing countries, where long-term global climate change may take a back seat to more pressing concerns such as clean water, a stable food supply, and economic growth.[7] In addition, countries differ significantly in the extent to which they are contributing to global climate change and the degree to which they will be impacted by it.

Even if international cooperation is achieved, there is no guarantee that governments, individually or collectively, will efficiently allocate climate resources. Government actions often are influenced by parties' engaging in what economists term *rent-seeking* behavior, attempting to redirect the economy's resources to their own advantage (e.g., ethanol mandates). The necessity of acting through voluntary international coalitions compounds this problem.

PUBLIC POLICIES TO COMBAT GREENHOUSE GAS EMISSIONS

Traditionally, U.S. environmental policy has relied on regulation and direct controls. A typical example is the use of mandates and incentives to switch to nonfossil fuel alternatives, such as renewable fuels standards (e.g., corn-based ethanol for vehicles) and renewable portfolio standards (e.g., wind power for electricity generation). Both are discussed in Chapter 8. This category also includes measures aimed at reducing energy consumption, such as mandated energy efficiency standards for home appliances and motor vehicle efficiency requirements for cars and trucks, often referred to as corporate average fuel economy (CAFE) standards. Many of these measures have been justified on a variety of grounds other than climate change, for example, energy security and air pollution control.

A recent decision by the EPA may greatly expand the use of regulation to reduce emissions of carbon dioxide. On December 7, 2009, the EPA declared that there was compelling scientific evidence that global warming from man-made greenhouse gases endangers Americans' health. The agency also determined that the pollutants (mainly carbon dioxide from burning fossil fuels) should be regulated under the Clean Air Act (EPA 2009b). These findings were in part a response to a 2007 U.S. Supreme Court decision that greenhouse gases fit within the Clean Air Act's definition of air pollutants. Although the findings do not in and of themselves impose any emission reduction requirements, they could pave the way for the government to require businesses that emit carbon dioxide and five other greenhouse gases to make costly changes in machinery to reduce emissions, even if Congress does not pass pending climate change legislation. They also allowed the EPA to finalize the greenhouse gas standards proposed earlier in 2009 for new light-duty vehicles as part of the joint rule-making with the Department of Transportation.

Direct controls are politically attractive because they do not explicitly convey the costs they impose. They are a form of hidden taxes. Unfortunately, this also means that they do not provide consumers and producers with information about the social costs of their decisions, making them less likely to be effective in mitigating global climate change (Nordhaus 2007; Parry and Pizer 2007). Research indicates that direct controls are also much more costly. For example, one study found that limiting U.S. greenhouse gas emissions through traditional regulatory approaches could be 10 times more costly than achieving the same result through a pricing policy (Burtraw et al. 2006). Research has also shown that gasoline and carbon taxes offer a much less expensive way to attain a given transportation emissions target than fuel economy standards (Austin and Dinan 2005; Aldy and Pizer 2008).

It has been observed that "emissions reductions are equally valuable wherever they occur, but they are not equally costly" (Kopp and Pizer 2007). A cost-effective climate change policy would seek to achieve emission reductions at the lowest possible cost. For this reason, economists generally favor addressing climate change through the use of incentive-based policies, either singly or in combination, that increase the cost of carbon emissions through fees or taxes, or restrict overall quantities of emissions (Yohe 1992; Newell and Stavins 2003; GAO 2008).

Setting a price on CO_2 emissions sends a clear signal to everyone engaged in activities that produce emissions (including both direct emitters and downstream consumers of the resulting products) about the social value of cutting emissions. Simultaneously, higher carbon prices provide incentives for research and development in low-carbon products and processes that can replace current technologies (Nordhaus 2007).

Almost a century ago, the English economist Arthur Pigou proposed taxing goods that were the source of a negative externality so as to reflect accurately the cost of the goods' production to society, thereby internalizing these costs (Pigou 1912). A tax on a negative externality, termed a Pigouvian tax, should equal the marginal cost of the damage. A number of European and Scandinavian countries have adopted some form of a carbon tax, and Costa Rica and Brazil are considering carbon taxes (Aldy and Pizer 2008). In the United States, the Clinton administration proposed, but was unsuccessful in getting enacted, a Btu tax.

The major advantage of a carbon tax is that it provides certainty as to the cost of using carbon, thereby facilitating long-term investments in carbon-saving technologies. The major drawbacks are that it may not limit atmospheric concentrations of greenhouse gases to the targeted levels and that the amount of abatement achieved will vary from year to year with the state of the economy (Parry and Pizer 2007).

An alternative approach to achieving emission reductions is through a cap-and-trade program. Entities that emit greenhouse gases are identified as the points of responsibility for emissions. Emission allowances (actually entries in an electronic bookkeeping system) are distributed such that the total is equal to the national cap, and covered entities must surrender allowances equal to their emissions or the emissions that result from their activities. Market trading in these allowances establishes a price on emissions that in turn creates economic incentives for cost-effective abatement (MIT 2007).

A cap-and-trade program ensures that the target concentrations will be reached, but the cost of achieving these levels is not known with certainty. A hybrid policy coupling a cap-and-trade program with an emissions price constraint (or safety valve) is garnering support among economists as providing both certainty and greater cost control (GAO 2008; Gayer 2009; Pew Center on Global Climate Change 2009c).

However the emission restrictions are achieved, they also would tend to reduce emissions of some conventional pollutants as well, yielding a variety of ancillary benefits, such as improvements in health from better air and water quality. Those additional benefits would partly offset the costs of greenhouse gas regulations (Congressional Budget Office 2003).

Economists also express support for mitigating climate change through public investment in invention, innovation, and education, which may be thought of as public goods as opposed to public bads, or negative externalities (GAO 2008). There is general agreement that new low-carbon technologies and processes will be required if climate change is to be effectively addressed. Through direct government funding or the use of tax credits, basic science as well as research and development can be encouraged.

THE KYOTO PROTOCOL: A GLOBAL APPROACH TO ADDRESSING CLIMATE CHANGE

International cooperation on the issue of climate change began with the creation of the Intergovernmental Panel on Climate Change (IPCC) in 1988 and came to fruition with the Kyoto Protocol (a draft treaty) in 1997 (see Chapter 8). The Kyoto Protocol is a cap-and-trade system that imposes national limits on the emissions of developed countries, requiring them on average to reduce their emissions 5.2% below their 1990 baseline over the 2008–2012 period. Developing nations are encouraged to reduce their greenhouse gas emissions but are not obligated to do so. The protocol created a framework and a set of rules for a global carbon market. It has resulted in the creation of several such markets, termed exchange trading schemes, with varying degrees of linkage among them.[8] It also led to the creation of a mechanism that regulated entities to use carbon credits (offsets) based on emission reduction projects.

The Fourth Assessment Report of the IPCC estimated mitigation costs for a range of greenhouse gas stabilization targets, ranging from 445 to 710 parts per million CO_2-equivalent in 2050 (IPCC 2007c). Based on a review of econometric models, the IPCC found that attaining these goals would produce a range of results, from a gain in GDP of 1% for the least ambitious target to a loss of 5.5% for the most ambitious goal, stabilizing global greenhouse gas concentrations below 535 parts per million (IPCC 2007c). The wide variability in results reflects the continued high level of uncertainty of both climate and econometric models.

These cost estimates are likely to be optimistic because the IPCC scenarios assume that mitigation (cap-and-trade) policies are introduced in a globally coordinated fashion such that the marginal cost of greenhouse gas abatement measures is equalized across all regions and countries (Peace and Weyant 2008). Any reduction or restriction in the number of participating countries or regions would increase both the carbon permit price and the economic cost associated with achieving a given stabilization target.

U.S. CLIMATE CHANGE POLICY

In addressing climate change, the United States has:

- implemented programs to support research and development and deployment of new technologies that reduce greenhouse gas emissions and that improve energy efficiency (e.g., the Department of Energy Climate Change Technology Program);
- invested in research to improve our understanding of climate change (e.g., the Climate Change Science Program); and
- introduced voluntary programs that provide technical assistance, education, and information-sharing (e.g., EPA's Climate Leaders Program) designed to encourage both public and private-sector entities to curb their greenhouse gas emissions (GAO 2008).

In addition, some programs that were intended to achieve other goals, such as pollution reduction, energy independence, and the limitation of soil erosion, also discourage emissions or encourage the removal of greenhouse gases from the atmosphere. Other programs, however, have opposing effects (Congressional Budget Office 2003).

A cap-and-trade approach has been employed to address the problem of power plant pollution that drifts from one state to another. Under the Acid Rain Program of the 1990 Clean Air Act, the EPA issued a Clean Air interstate rule covering 28 eastern states and the District of Columbia. It seeks to use a cap-and-trade system to reduce the target pollutants, sulfur dioxide (SO_2) and nitrogen oxides (NOx) by 70% (Chesnut and Mills 2005).

Since 1990, SO_2 emissions have dropped 59% (calculated from data in EPA 2009a). The EPA has estimated that by 2010, the overall costs of complying with the program for businesses and consumers will be \$1–\$2 billion a year, only one fourth of what was originally predicted (EPA 2006).[9]

States and regions have developed their own initiatives to restrict CO_2 emissions (Litz 2008; Pew Center on Global Climate Change 2009a). In 2006, California became the first to enact such a comprehensive statute. Through similar legislation, Hawaii, Minnesota, New Jersey, Oregon, and Washington also have established greenhouse gas emissions targets. Governors in many other states have issued executive orders or plans setting statewide, economywide greenhouse gas reduction targets (see Chapter 8). Although for the most part these orders and plans do not have the full force of state law, they provide the impetus for significant action to reduce emissions. At the regional level, in 2003, New York State attained commitments from nine other northeastern states to form a cap-and-trade program for CO_2 emissions from power generators (Regional Greenhouse Gas Initiative 2003). Emission permit auctioning began in September 2008, and the first three-year compliance period began on January 1, 2009. The aim is to reduce the carbon budget of each state's electricity generation sector to 10% below the 2009 allowances by 2018.[10]

These actions at the state and regional levels are already causing companies to view climate change as a serious business issue. They also create uncertainty and the potential for firms to suffer, if their competitors located in other areas do not face such constraints. This increases the political pressure for nationwide action to reduce carbon emissions.

National Policy

It appears increasingly likely that the United States will enact some type of comprehensive policy to limit future emissions of CO_2.

Three proposals introduced in the 111th Congress would use a carbon tax approach to address CO_2 emissions from fossil fuel combustion: H.R. 594 (Stark), H.R. 1337 (Larson), and H.R. 2380 (Inglis). Another, H.R. 1683 (McDermott), would establish a program that also may be described as a dynamic carbon tax; its tax rate would be linked with annual emission allocations (or caps). The Stark and McDermott proposals seek to combine near-term price certainty with specific emissions targets, calling for a tax that rises continually to ensure that the nation's CO_2 emissions will fall to 80% below 1990 values by 2050.

The only climate change legislation to pass one house of Congress is the American Clean Energy and Security (ACES) Act, passed by the House of

Representatives on June 26, 2009, by a vote of 219–212. The legislation creates a mandatory economywide cap-and-trade program to regulate emissions of greenhouse gases, including carbon dioxide, methane, nitrous oxide, sulfur hexafluoride, perfluorocarbons, nitrogen trifluoride, and certain hydrocarbons. The ACES Act also requires up to 20% of electricity production to be from renewable sources; increases energy efficiency standards for buildings, appliances, and industry; increases investment in renewable energy, carbon capture, and sequestration technology; and provides assistance to low-income families and workers impacted by the provisions of the bill (Congressional Budget Office 2009a).

The U.S. Senate is considering three cap-and-trade bills:

- The Kerry-Boxer Clean Energy Jobs and American Power Act (S. 1733) has been passed by the U.S. Senate Environment and Public Works Committee. In most respects, this bill is similar in approach and content to H.R. 2454 (Resources for the Future 2010; Pew Center on Global Climate Change 2010).

- The Cantwell-Collins Carbon Limits and Energy for American Renewal Act (S. 2877) also imposes an economywide cap on emissions of carbon dioxide, but unlike H.R. 2454 and S. 1733, it requires 75% of auction proceeds to be distributed to U.S. citizens and 25% to be placed in the Clean Energy Reinvestment Trust Fund (Pew Center on Global Climate Change 2010; Boyce and Riddle 2010).

- The Bingham American Clean Energy Leadership Act (S. 1462), passed out of the Senate Committee on Energy and Natural Resources on a bipartisan vote of 15–8, takes a different approach. It was passed as an energy bill with provisions related to increased energy production, energy efficiency, renewable energy standards, technology research and development, energy market stabilization, and transmission network improvements, among others (Pew Center on Global Climate Change 2009b).

The Impacts of a Cap-and-Trade Approach

A number of public and privately sponsored studies have assessed the potential economic impacts of U.S. cap-and-trade climate legislation (MIT 2007, 2008, 2009; CRA International 2007; Murray and Ross 2007; EIA 2008c, 2009; ICF International 2007, 2008; Kopp and Pizer 2007; Aldy and Pizer 2008; EPA 2008, 2009c; Charles River Associates 2009; Heritage Center for

Data Analysis 2009; Science Applications International 2009; Congressional Budget Office 2009a-c). These studies differ in terms of their focus, assumptions, and time frames (Parker and Yacobucci 2009). Nonetheless, it is possible to draw some general conclusions about the likely economic impacts of a national cap-and-trade policy.

Coverage and Scope

More than 80% of greenhouse gas emissions occur as a byproduct of the combustion of fossil fuel (coal, oil, and natural gas), with the remaining 20% coming from fugitive emissions of nitrous oxides and methane from agriculture and industrial releases of fluorinated gases and nitrous oxides (EPA 2008). Economic research suggests that a cost-effective cap-and-trade program should exploit emission abatement opportunities among as many sources as possible.[11] Although pending bills address primarily the emissions that come from fossil fuel combustion, they nonetheless cover 80–85% or more of all emissions (termed economywide controls).

Although it is possible to design a cap-and-trade program to be applied downstream (e.g., to residences or gasoline stations), most of the pending bills focus entirely or primarily on upstream regulation (e.g., at the mouth of a mine or refinery gate). Regulation at or near the point of fossil fuel production would involve modest monitoring costs, since only 2,000–3,000 facilities would need to be covered to control all fossil fuel CO_2 emissions (Stavins 2007; Hall 2007). It should be noted that even if the focus of the program is upstream, it is estimated that 90% or more of the higher carbon price would be passed forward to consumers (Lasky 2003).

Emission Reductions

Pending cap-and-trade proposals are projected to result in emission reductions of 20% by 2020, 42% by 2030, and 83% by 2050 (Resources for the Future 2010). The basic assumption underlying cap-and-trade proposals is that the setting of stringent targets starting in 2012 will cause utilities to switch from coal to other fuels, especially natural gas, sooner rather than later. If, as is commonly assumed, clean coal and carbon capture and storage technologies become commercially viable after 2025, utilities are expected to turn once more to fossil fuels; but with reduced demand and carbon capture and storage in place, CO_2 emissions will continue to fall. It is forecast that by mid-century (2045) 90% of CO_2 emissions from the electricity sector will be captured through these technologies.[12] Thereafter additional reductions must come from other sectors, where abatement comes at a higher cost.

Impact on Prices and Output

Consumer prices will rise once the greenhouse gas allowance price is incorporated, but supply (or producer) prices may fall in response to reduced demand. Some studies project large increases in electricity prices as utilities transition to decarbonized generation. The failure to include the effect of consumer rebates in the initial years, however, leads to an overstatement of the increase. If consumer rebates are taken into consideration, the projected impacts on electricity rates by 2020 is small (a rise of 13% or less). Electricity rates are projected to escalate thereafter, when the rebates have been phased out, resulting in increases averaging about 20% in 2030 and as much as 49% in 2050 (Parker and Yacobucci 2009).

The costs of a cap-and-trade program by 2050, when compared with 2005, generally are estimated to be 1–2% of national income, roughly comparable to the estimated costs of all other environmental policies combined (Aldy and Pizer 2008; MIT 2007, 2008; EPA 2008; Parker and Yacobucci 2009). More stringent emissions targets or cost-ineffective policy implementation could substantially increase abatement costs.

Given the dominant position of Texas as an energy-intensive state, it might be expected that cap-and-trade legislation would place more of an economic burden on Texas than on other states. A study conducted by the Center for Energy Economics in the Bureau of Economic Geology at the University of Texas at Austin, at the request of the Texas Comptroller of Public Accounts, projects that under H.R. 2454, by 2030 total employment in Texas declines by just over 1.1% (about 137,000 jobs lost). Gross state product falls by almost 1.7%, almost $26 billion in constant 2000 dollars, and real disposable income falls roughly 0.9%, or about $24 billion in constant 2000 dollars (Foss and Gulen 2009).

Cost Containment

The uncertainty surrounding the costs of implementing a cap-and-trade program has generated interest in including mechanisms to contain these costs. Cost containment is likely to be more important in the short run, as the economy adjusts to carbon constraints, than in the longer run, when broader global efforts to stabilize greenhouse gas atmospheric concentrations at a specific level may take precedence (Kopp and Pizer 2007).

Some proposals allow regulated entities to shift their obligations across periods, banking emission units for future use or borrowing by shifting deficits forward. This can improve near-term cost certainty by limiting allowance price volatility. These provisions are usually phased out over time.

Another approach to cost containment would establish a price cap, termed a safety-valve provision; the government would agree to sell allowances in unlimited amounts at a fixed price to guard against price spikes that may sharply raise compliance costs or threaten the survival of the system itself. Other legislation seeks to contain costs by including specific limits on allowance prices, which gradually increase over time.

One feature of a cap-and-trade program that potentially can limit costs is to permit offsets under which projects that reduce greenhouse gas emissions relative to some agreed baseline are granted credits equal to the volume of reductions. These credits can be sold into a cap-and-trade program. Linking a cap-and-trade program to outside systems has two advantages. First, it enables domestic producers to use offsets to take advantage of the relatively low-cost emission reductions available in other countries or in domestic sectors (e.g., agriculture) that are not covered by the program. Second, it offers a cost-effective means of achieving emission reductions in developing countries where a full-fledged emissions trading program would be unlikely because of political and institutional constraints (Hall 2007). Such a system faces a host of difficulties, not the least of which are high transaction costs due to the need to validate and monitor the effectiveness of the offset projects. Both the Waxman-Markey (H.R. 2454) and Kerry-Boxer (S. 1733) proposals contain provisions permitting the use of domestic offsets and international credits to achieve their net emission reduction targets. The offsets and credits, however, are gradually reduced over time. On the other hand, the Cantwell-Collins (S. 877) proposal does not provide for market-based offsets (Resources for the Future 2010).

Although the Waxman-Markey proposal does not contain a safety valve, it does address cost control through five main mechanisms: (1) unlimited banking and borrowing, (2) a two-year compliance period, (3) a strategic reserve auction with a pool of allowances available at a minimum reserve price, (4) periodic auctions with a reserve price, and (5) broad limits restricting the use of offsets to two tons of emissions annually, which have the potential to reduce the projected allowance prices by half or more (Parker and Yacobucci 2009). The Kerry-Boxer proposal contains similar cost-containment provisions. The Cantwell-Collins proposal does contain a safety valve, in the form of a collar set with a floor of $7 and a ceiling of $21 in 2012, with the floor and ceiling both rising over time (Resources for the Future 2010).

Allowance Allocation

The primary mechanism that any market-based system uses to control greenhouse gas emissions is a price signal. Under a cap-and-trade program, capping emissions creates a new commodity, an allowance (essentially a license to emit one ton of CO_2 equivalent). Supply and demand determine the allowance price, and the market responds by efficiently reducing carbon emissions to the point where the cost of lowering emissions (compliance cost) equals the cost of buying an allowance. By creating allowances, the government is creating an instrument that has value and for which there will be a demand at a given price. Models analyzing the economic impacts of H.R. 2454 project allowance prices between $13 and $21 in 2015, which thereafter rise 4–6% annually. The total value distributed each year could approach $100 billion by 2040, as projected allowance prices (in constant 2005 dollars) rise from $10.80 in 2012 to $42.33 in 2040 (Parker and Yacobucci 2009). The availability of offsets significantly lowers the cost of the program; their absence would likely increase these costs.

EFFICIENCY-EQUITY TRADE-OFF

A public policy can be evaluated in terms of its efficiency. In simple terms, economic efficiency is achieved when a given result is achieved at the lowest possible cost. Equity refers to the distributional burden across households. If the costs imposed by a public policy fall disproportionately on lower-income households, it is said to be regressive. If the costs are borne disproportionately by higher-income households, the policy is said to be progressive. In many instances, equity and efficiency are trade-offs in the sense that an improvement in efficiency may come at the expense of a reduction in equity.

Implementation of a cap-and-trade program raises the price of fossil fuels and products that are made from fossil fuels, such as electricity. Consumers who were willing to pay to buy these products at the initial price now find that they either have to pay more or consume less. This results in a decrease in consumer surplus and a decline in economic welfare.

Changes in consumer surplus, however, do not account for ancillary effects from changes in employment and income that also occur when the price of a product rises. A cap-and-trade program will shift economic activity away from relatively energy-intensive sectors of the economy to those that are less energy-intensive. This shift likely would lead to unemployment for displaced workers and may force some workers to accept jobs with lower wages. To the extent that lower-wage workers are employed by energy-intensive industries or in regions (such as Texas) that would experience a reduction in economic activity, these employment and income aspects could be regressive. Taking into account the included cost-containment provisions, and assuming that much of the revenue raised would be returned in some form to households, recent studies suggest that the impact on households of cap-and-trade legislation would be fairly small, less than $200 per year (Pew Center on Global Climate Change 2009c; Congressional Budget Office 2009c).

A recent study evaluating the short-term (to 2015) effects of a national cap-and-trade program on households in 11 regions, one of which is Texas, finds that the incidence (the distribution of costs) depends critically on how the program distributes the value created by placing a price on carbon dioxide emissions (Burtraw et al. 2008).[13]

For both Texas and the nation, free allocation of allowances provides neither equity nor efficiency gains. The largest gains in equity come from using allowance sales revenues to expand the earned income tax credit. These gains come at the cost of a modest loss in efficiency. Using revenues from allowance sales to invest in improving efficiency offers strong equity and efficiency gains.

COST-BENEFIT ANALYSIS

In developing and assessing public policies, economists prefer to balance the costs and benefits of proposed actions. It is much more difficult, however, to estimate the economic impacts of projected climate change than it is to estimate the cost of policies to avoid such impacts. This is because the effects of climate change are location-specific and are generally thought to occur gradually over long periods of time (i.e., centuries). At the national and global levels, climate change models yield a wide range of possible temperature and other climate changes that are likely to occur over broad geographic areas that generally do not reflect political boundaries. For local and regional analysis, there is a dearth of data at the level of detail needed to make an accurate impact assessment. Since there are sizable differences between economic sectors in terms of their vulnerability to climate change, comprehensive integrative models are needed to assess impacts in terms of the entire economy. Three additional problems make it especially difficult to estimate the present value of damages from projected climate change.

- Projected economic damages from climate change are quite modest initially, begin to become somewhat significant at mid-century, and increase in severity as we reach the end of the century and beyond. When the longer-term damages are discounted to the present in a cost-benefit analysis, they become much less significant (Sheraga and Sussman 1998; Nordhaus 2007).[14]

- There is a high degree of scientific uncertainty with regard to the magnitude of climate change (Heal and Kriström 2002; Newell and Pizer 2001). To complicate the analysis further, one or more possible, highly undesirable outcomes from climate change may be irreversible, and the possibility exists for abrupt and catastrophic changes in climate (IPCC 2007a). In light of the variability surrounding future climate paths, it remains difficult to assign probability estimates to possible climate scenarios.[15]
- There are many impacts of climate change that cannot be easily expressed in monetary terms (Howarth and Monahan 1996). Efforts by economists to find ways of valuing non-market resources have not been successful in garnering widespread acceptance.

Integrated assessment models project both the costs of policies to limit greenhouse gas emissions and the benefits of such action (largely defined in terms of avoided damages). One of the most prominent integrated assessments uses the DICE model developed by William Nordhaus (Nordhaus 1994; Nordhaus and Boyer 2000). The latest projections using this model indicate that the net present value of the global benefit of an optimal policy to reduce greenhouse gas emissions is $3 trillion, relative to no controls (Nordhaus 2007). This total involves $2 trillion in projected abatement costs and $5 trillion of reduced climatic damages. An estimated $17 trillion in damages would remain, but additional abatement would cost more than the additional reduction in damages (which is considered nonoptimal).

In 2008, the U.S. General Accountability Office (GAO) convened a panel of experts to consider the economics of policy options to address climate change (GAO 2008). All of the 18 panelists agreed that Congress should consider using a market-based mechanism to establish a price on greenhouse gas emissions. Most panelists preferred either a tax on emissions or a hybrid policy that incorporates a cap-and-trade program with a safety-valve provision, under which the government would sell additional emission permits if the permit price rose above a specified level. A majority recommended additional actions as part of a portfolio to address climate change, including investment in research and development of low-emission technologies.

The GAO noted that 14 of the 18 panelists were at least "moderately certain" that the benefits of their recommended portfolio of actions would outweigh the costs. The panel rated estimates of costs as more useful than estimates of benefits for informing congressional decision making. In other words, the panelists rejected making the decision to limit carbon emissions on a cost-benefit basis (reflecting concerns about the major issues discussed above in determining benefits). Instead, they emphasized the role of economics in helping determine the least-cost means of accomplishing whatever level of carbon emissions is established.

CAP-AND-TRADE INCREASES THE IMPORTANCE OF NONFOSSIL FUEL ALTERNATIVES

A number of engineering studies and other analyses have identified ample opportunities to improve energy efficiency at relatively low cost. From an economic perspective, however, such opportunities would exist only if there were persistent market failures. Although these cannot be completely ruled out, there is a good deal of evidence that consumers and firms respond relatively quickly to perceived economic opportunities. For example, as oil prices soared in 2008, sales of fuel-inefficient cars plunged and transit ridership rose (*Financial Times* 2008). A higher price for carbon would likely spur efforts to improve energy efficiency.

Beginning with the 1978 Energy Tax Act, the United States has used financial incentives to promote renewable energy. The Energy Policy Act (EPACT) of 1992 established a 10-year production tax credit of 15 cents per kilowatt hour, adjusted for inflation, for tax-paying privately and investor-owned wind projects and closed-loop biomass plants.[16] EPACT also created the Renewable Energy Production Incentive (REPI) for electricity generated from biomass, geothermal, wind, and solar by tax-exempt publicly owned utilities and rural cooperatives. Unlike the production credit, the funding available through REPI was subject to annual congressional appropriations, thereby making the availability and level of the credit uncertain.[17]

Carbon dioxide–emitting fossil fuels accounted for about 85% of total U.S. energy consumption in 2006. The remainder was divided between nuclear electric power (8%) and renewable energy (7%). Most of the renewable energy came from hydroelectric power (42%), wood (31%), and biofuels (11%), with solar-photovoltaic (1%), wind (4%), geothermal (5%), and waste (6%) making up the remainder (EIA 2008b). Projections by the federal Energy Information Administration for 2030 indicate that renewable energy incentive programs are expected to have a modest impact. The fossil fuel share of consumption is projected to decline from 85% in 2006 to 82.4% in 2030, while the nonfossil fuel share rises to 17.6%. Within nonfossil fuels, the share of consumption accounted for by nuclear power is expected to remain relatively constant, while the share from hydropower drops 2.9–2.5%. Most of the increase in nonfossil fuel consumption occurs in biomass, which jumps from 2.5% to 4.7%. Other renewable energy sources (including wind and solar) are projected to increase 4.4% annually, but they are projected to provide only about 2.1% of total energy consumption in 2030.

Since the introduction of a cap-and-trade policy substantially increases the cost of using fossil fuels, it might be expected that it would result in an increase in nonfossil fuels as a share of total U.S. energy consumption. The cap-and-trade program would primarily impact the fuel mix used in generating electricity, since it is expected to have only a minor impact on the transportation sector (about one third of carbon dioxide emissions).[18]

Electricity consumption, including both purchases from electric power producers and onsite generation, is projected to increase at an average annual rate of 1.15% between 2006 and 2050 (EIA 2008a).[19] The share of electricity generated by CO_2-emitting fossil fuels steadily rises, while the share of nuclear and other renewable resources declines. Adoption of the proposed cap-and-trade program would reverse these trends. Although natural gas is substituted for coal in the early years (to 2030), coal regains market share in the later years (2030–2050), after the introduction of commercially viable carbon capture and storage (CCS) technology. Nonetheless, the share of electricity generation from fossil fuels is forecast to fall to 53.6% by 2030, as the share of nuclear energy rises to nearly 30% and other nonfossil sources climb to 16.6%. Even with carbon sequestration, the share of fossil fuels in electricity generation is expected to continue to decline, dropping to 38.5% by 2050 as the shares of nuclear and other renewables climb to 36.6% and 24.9%, respectively.

There are many uncertainties that affect the impacts of proposed cap-and-trade programs. These include the degree to which new nuclear power and biomass projects are technically, socially, and politically feasible and whether commercially viable CCS technology will be available on a large scale (EPA 2008). Under assumptions that limit the growth of nuclear, biomass, and CCS technologies, electricity prices are projected to be 79% higher in 2030 and 98% higher in 2050 than under the reference scenario. This would spur development of electricity generation from renewable resources. Higher prices would also reduce the overall demand for electricity, projected to be 7% lower in 2030 and 14% lower in 2050 than under the cap-and-trade program with unconstrained nuclear, biomass, and CCS technology. As a result, the share of renewable resources would actually decline to 11.6% in 2030 and 10.5% in 2050, compared with the cap-and-trade program projections with no constraints (16.6% and 24.9%).

One study estimates that it would be technically feasible to rely on wind energy to generate one fifth of U.S. electrical energy (DOE 2008). The primary stumbling block would be the huge investment ($60 billion between now and 2030 under optimistic assumptions) required to reconfigure the nation's electric grid to bring wind energy generated in remote areas to urban markets. Another obstacle is the variability of the availability of wind power over the 24-hour day and seasonally. Finally, large wind power and transmission projects may face serious political and environmental opposition. These obstacles are more likely to be overcome under a cap-and-trade program with constrained development of nuclear, biomass, and CCS technology.

A recent report emphasizes the potential of solar energy, long derided as being uneconomical (Lorenz et al. 2008). Over the last 20 years, the cost of manufacturing and installing a photovoltaic solar power system has fallen 20% with every doubling of installed capacity. Meanwhile, the cost of generating electricity from conventional sources, such as natural gas, has been rising. As a result, solar power has been edging closer to being cost-competitive in some markets. Within three to seven years, it is projected that unsubsidized solar power, in markets such as California and Texas, could cost no more to end customers than electricity generated by fossil fuels. It is estimated that by 2020, global installed solar capacity might be 20–40 times its level today and account for 3–6% of installed generation capacity (1.5–3% of output). As with wind energy, solar power must overcome a number of obstacles to gain market share. Foremost among them are choosing among competing technologies, and developing appropriate regulatory policies to provide incentives that encourage production in the short run but that phase out as solar power reaches grid parity with other power generation sources.

Texas may have many opportunities to develop new industries under a cap-and-trade program. The Lone Star State leads the nation in nonhydro-power renewable energy potential, with a large amount of wind generation capacity and high levels of direct solar radiation capable of supporting solar power generation (EIA 2008b).[20] Thanks to its large agricultural and forestry sectors, Texas also has an abundance of biomass energy resources. Texas already is a major nuclear power generating state. Two nuclear plants, Comanche Peak and the South Texas Project, are among the largest in the nation, accounting for nearly one eighth of total net electricity generation.

Although renewable energy sources contribute minimally to the state's power grid, accounting for only 3% of total net electricity generation, Texas leads the nation in wind-powered generation capacity, with 27% of the U.S.

total. Improvements in wind generation technology and reductions in development costs are leading to construction of substantial new wind generation capacity (Texas Comptroller 2008). The state's pulp and paper industry uses the biomass energy from wood to generate the electricity, heat, and steam it uses onsite. Texas also is the largest producer of biodiesel transportation fuel in the United States, but it is only a minor producer of ethanol (EIA 2008b).

CONCLUSIONS

Projected climate change over the next century is likely to have only a small measurable impact on the Texas economy. Many of the potential impacts involve problems that are already recognized. Texas needs to address issues such as sea-level rise and coastal erosion, air quality in its major urban centers, overreliance on coal for electricity generation, dwindling availability of water (including the conflict between agricultural and urban use), and water quality. Effectively addressing these issues will go a long way toward mitigating the impact of climate change on the state.

It appears increasingly likely that the nation will adopt some form of cap-and-trade legislation to reduce greenhouse gas emissions. If that happens, market forces will bring about significant changes in the state's energy mix. The average economic impact of these changes, as well as the impact on the state's poorest households, will depend to a significant degree on the specifics of the policy, especially on whether the allowances are distributed without charge or auctioned off. In addition, cap-and-trade policies that permit the use of domestic and international offsets, allow banking and borrowing of allowances, provide assistance for transition to emission-saving technologies, and contain a safety valve to limit costs will reduce, but not eliminate, the relatively higher economic burden placed on the state.

Texas also is in a unique position to benefit from expanded use of non-fossil fuel sources for electricity production. The nuclear power industry in the state is well established. Although Texas is leading the way in converting wind energy into electricity, much untapped potential remains. The state is also rich in solar and biomass resources.

Looking to mid-century, it is clear that the cost to Texas of a national cap-and-trade policy would likely exceed any possible measurable benefit in terms of avoided damages. Over a longer time frame, if the harmful impacts of climate change continue to increase, the cost-benefit balance might shift. Time is not on our side. Texas would benefit economically by taking stronger actions today to address climate change impacts at the state level, and by supporting the adoption of cost-effective, equitable policies at the national level to limit greenhouse gas emissions and encourage the use of nonfossil fuel alternatives.

NOTES

1. This is a marked departure from the pattern of growth experienced in 2000–2004. If those years are used as the basis for projecting future population, 116 counties, including rural counties and some metropolitan counties—especially in West Texas—would lose population, leading to further concentration of population in suburban and some central city counties. Central cities generally consist of one or more of the largest population and employment centers of a metropolitan area (e.g., San Antonio in Bexar County and Austin in Travis County).

2. Daily loads are U-shaped, reaching a minimum at about 65°F. The hot-side slopes are similar across the regions, with a range of 0.82–1.05 kilowatt hour (kWh) per °F. These slopes are amplified above 80°F (by an additive slope of 0.57 kWh per °F) but are muted beyond 85°F (by an additive slope of –0.15 kWh per °F). Cold-side slopes across regions are also strong, with a range of 0.66–1.39 kWh per °F below 60°F. These slopes are amplified below 50°F by an additive slope of 1.09 kWh per degree.

3. The projected increase in peak demand relates not only to average increases in temperature but also to projected increases in the incidence of extreme heat events caused by climate change.

4. Getting the additional power to market would also require significant investment in new transmission capacity. For example, a recent study completed by ERCOT on the potential costs to build transmission lines to West and Northwest Texas to transport electricity generated from wind power estimated that it would cost $3–$6 billion, depending on the amount and capacity of transmission lines built (ERCOT 2008).

5. The Constanza study examined the economic value of 16 biomes, or ecosystem types, and 17 of their key goods and services, including nutrient cycling, disturbance regulation, waste treatment, food production, raw materials, refuges for commercially and recreationally important species, genetic resources, and opportunities for recreational and cultural activities.

6. A ton of CO_2 makes the same contribution to climate change regardless of the location of emissions in the world. Emissions of other greenhouse gases will generally have different radiative forcings and different atmospheric lifetimes; however, their global warming potential can be

converted to equivalent CO_2 units, conventionally expressed as metric tons of CO_2e (IPCC 2001). One metric ton is equal to 1,000 kilograms or 1.1 short tons (2,000 pounds).

7. A recent example of this trade-off is the proposal by Eskom, South Africa's energy utility, to build a 4,800 megawatt coal-fired power station that would emit an estimated 25 million tons of CO_2 each year. It is being justified on the grounds that it would help solve a shortage of generating capacity that threatens to undermine the economic growth of South Africa (Lomborg 2010).

8. The European Union Emission Trading Scheme is the largest such multinational allowances program in the world. It currently is the world's only mandatory carbon trading program. Phase 1 (2005–2007) began in January 2005, with 25 of the 27 European Union countries participating. Phase 2 (2005–2012) links the European scheme to other countries participating in the Kyoto trading system. It is estimated that worldwide global trade in emission allowances reached $56 billion in 2007 and then declined to $30 billion in 2008 (Sjardin 2008).

9. The future of the SO_2 cap-and-trade policy was cast into doubt by a decision by the U.S. Court of Appeals for the D.C. Circuit on July 11, 2008, which invalidated the regulation. The effect was potentially to make worthless $15–$20 billion in allowances. The EPA and other parties requested a rehearing, and on December 23, 2008, the court revised its decision and remanded the rule to the EPA without annulling it. Thus, the rule remains in place until the EPA issues new rules to replace it. The resulting uncertainty highlights one problem with regulated solutions to problems: the risk that the courts or a future Congress will change the rules of the game (Dizard 2008).

10. The number of permits auctioned off for 2009 actually allowed more emissions than the plants were expected to emit. Emissions have been dropping, in part because power plants have switched to natural gas as its price has fallen (Ball 2008). For an excellent analysis of why models cannot reliably predict the future costs of climate change, see Parker and Yacobucci (2009).

11. One study concluded that compliance costs were cut in half when gases other than CO_2 were included in mitigation policies (Weyant et al. 2006). Studies of U.S. mitigation costs reached similar conclusions (EIA 2009).

12. The electricity sector is responsible for 40% of our CO_2 emissions, but it is expected to provide two thirds to three quarters of the emission

reductions that would be achieved during the first couple of decades of a cap-and-trade program (Burtraw 2008).

13. The analysis would be equally valid for distributing the revenue from a carbon tax.

14. The Nordhaus study imposed a discount rate that gradually declined from over 4% today to under 3% in 100 years. This led the model to assign a present value of about $25 billion (one fortieth of the future value) to $1 trillion of damages a century from now. Some economists support a much lower discount rate for assessing long-term damages (Cline 1992).

15. The range of possible outcomes in terms of temperature and precipitation cannot be expressed in a single number, nor does averaging them help. For example, assume that 99 times out of 100, the right number is $1.00, but 1 in 100 times the right number is $51.00. The average, $1.50, is not a useful representation of the actual situation. In cost-benefit terms, one would not be willing to spend $2 to avoid a $1.50 outcome, but one might be willing to spend much more than $2 to ensure that the $51 outcome does not occur.

16. EPACT significantly improved the economics of wind power, but net metering programs, implemented at the state level, have been more beneficial to the installation of solar photovoltaic generation. These programs are designed for small electricity customers (residential or small commercial) who produce their own power, allowing them to bank power on the grid in times of surplus and draw down from the grid in times of need (Texas Comptroller 2008).

17. The incentive expired in 1999 but was renewed twice, in 1999 and 2001, before its expiration at the end of 2003. Late in 2004, it was extended again through 2005. This latest extension increased the number of renewable technologies that are covered by the incentive.

18. This is because of the relatively modest indirect price signal an upstream cap-and-trade program sends to the transportation sector. The price signal provided by Senate bill 2191 (an increase in the price of gasoline of about $0.53 per gallon in 2030 and about $1.40 in 2050) is not high enough to cause large changes in the demand for transportation or changes in how transportation services are provided (EPA 2008).

19. For comparison, electricity consumption grew by annual rates of 4.2% in the 1970s, 2.6% in the 1980s, and 2.3% in the 1990s.

20. Although Texas is not known as a major hydropower state, substantial untapped potential exists in several river basins, including the Colorado

River of Texas and the Lower Red River (EIA 2008b). The importance of managing Texas water as a scarce resource is likely to outweigh the relatively tiny amount of power it could add to the state's electricity grid (Texas Comptroller 2008).

REFERENCES

Aldy, J. E., 2007. Divergence in State-Level Per Capita Carbon Dioxide Emissions. *Land Economics* 83(3):353–369.

Aldy, J. E., and W. A. Pizer, 2008. *Issues in Designing U.S. Climate Change Policy.* RFF Discussion Paper 08–20. Resources for the Future, Washington, D.C.

Austin, D., and T. Dinan, 2005. Clearing the Air: The Costs and Consequences of Higher CAFE Standards and Increased Gasoline Taxes. *Journal of Environmental Economics and Management* 50:562–582.

Ball, J., 2008. Efforts to Curtail Emissions Gain. *Wall Street Journal,* September 22, 2008.

Berg, R., 2009. *Tropical Cyclone Report: Hurricane Ike.* Report AL092008, as amended February 4, 2009. National Hurricane Center, Miami, Fla.

Boyce, J. K., and M. E. Riddle, 2010. *Clear Economics: State-Level Impacts of the Carbon Limits and Energy for America's Renewal Act on Family Incomes and Jobs.* Political Economy Research Institute, Department of Economics, University of Massachusetts, Amherst.

Burtraw, D., 2008. Preventing Climate Change: Second in a Series of Hearings. Testimony prepared for the U.S. House of Representatives Committee on Ways and Means, September 18, 2008. Resources for the Future, Washington, D.C.

Burtraw, D., W. Pizer, W. Harrington, J. Sanchirico, and R. Newell, 2006. Modeling Economywide versus Sectoral Climate Policies Using Combined Aggregate-Sectoral Models. *Energy Journal* 27(3):135–168.

Burtraw, D., R. Sweeney, and M. A. Walls, 2008. *The Incidence of U.S. Climate Policy: Where You Stand Depends on Where You Sit.* RFF Discussion Paper 08–28. Resources for the Future, Washington, D.C.

Charles River Associates, 2009. *Impact on the Economy of the American Clean Energy and Security Act of 2009 (H.R. 2454).* Prepared for the National Black Chamber of Commerce, Washington, D.C.

Chesnut, L. G., and D. M. Mills, 2005. A Fresh Look at the Benefits and Costs of the U.S. Acid Rain Program. *Journal of Environmental Management* 77:225–266.

Climate Change Science Program, 2007. *Effects of Climate Change on Energy Production and Use in the United States.* Synthesis and Assessment Product 4.5. National Science and Technology Council, U.S. Climate Change Science Program, Washington, D.C.

Cline, W. R., 1992. *The Economics of Global Warming.* Institute for International Economics, Washington, D.C.

Congressional Budget Office, 2003. *The Economics of Climate Change: A Primer.* Congressional Budget Office, Washington, D.C.

Congressional Budget Office, 2009a. *Cost Estimate H.R. 2454 American Clean Energy and Security Act of 2009*. Congressional Budget Office, Washington, D.C.

Congressional Budget Office, 2009b. *The Economic Effects of Legislation to Reduce Greenhouse Gas Emissions*. Congressional Budget Office, Washington, D.C.

Congressional Budget Office, 2009c. *The Estimated Costs to Households from the Cap-and-Trade Provisions of H.R. 2454*. Congressional Budget Office, Washington, D.C.

Constanza, R., R. d'Arge, R. de Groot, S. Farberparallel, M. Grasso, B. Hannon, K. Limburg, S. Naeem, R. V. O'Neill, J. Paruelo, R. G. Raskin, P. Sutton, and M. van den Belt, 1997. The Value of the World's Ecosystem Services and Natural Capital. *Nature* 387:253–260.

CRA International, 2007. *Economic Impacts of Proposed House/Senate Legislation on the U.S. Economy*. CRA International, Washington, D.C.

Dizard, J., 2008. Misguided Game of Sox and NOx Played for High Stakes. *Financial Times*, August 19, 2008.

DOE, 2008. *20% Wind Energy by 2030*. DOE/GO-102008–2567, May 2008. U.S. Department of Energy, Washington, D.C.

EEA, 2005. *Hurricane Damage to Natural Gas Infrastructure and Its Effect on U.S. Natural Gas Market*. Energy and Environmental Analysis, Inc., Arlington, Va.

EEI, 2005. *After the Disaster: Utility Restoration Cost Recovery*. Edison Electric Institute, Washington, D.C.

EIA, 2008a. *Annual Energy Outlook*. U.S. Energy Information Administration, Washington, D.C.

EIA, 2008b. *State Energy Profiles: Texas*. U.S. Energy Information Administration, Washington, D.C.

EIA, 2008c. *Energy Market and Economic Impacts of S2191, the Lieberman/Warner Climate Security Act of 2008*. U.S. Energy Information Administration, Washington, D.C.

EIA, 2009. *Energy Market and Economic Impacts of H.R. 2454, the American Clean Energy and Security Act of 2009*. Energy Information Administration, Washington, D.C.

Emanuel, K., 2005. Increasing Destructiveness of Tropical Cyclones over the Past 30 Years. *Nature* 436:686–688.

EPA, 2006. *Acid Rain Progress Report 2006*. U.S. Environmental Protection Agency, Washington, D.C.

EPA, 2008. *Analysis of the Lieberman-Warner Climate Security Act of 2008 S. 2191 in 110th Congress*. U.S. Environmental Protection Agency, Washington, D.C.

EPA, 2009a. *Sulfur Dioxide: Air Trends*. U.S. Environmental Protection Agency, Washington, D.C.

EPA, 2009b. *EPA: Greenhouse Gases Threaten Public Health and the Environment*. News Release, Office of Air and Radiation. U.S. Environmental Protection Agency, Washington, D.C.

EPA, 2009c. *EPA Analysis of the American Clean Energy and Security Act of 2009: H.R. 2454 in the 111th Congress*. U.S. Environmental Protection Agency, Washington, D.C.

EPRI, 2003: *A Survey of Water Use and Sustainability in the United States with a Focus on Power Generation*. Report 1005474. Electric Power Research Institute, Washington, D.C.

ERCOT, 2006. *Capacity, Demand, Reserve Report*. Electric Reliability Council of Texas, Austin.

ERCOT, 2008. *ERCOT's Competitive Renewable Energy Zone Transmission Optimization Study*. Electric Reliability Council of Texas, Austin.

Financial Times, 2008. Strategic Choice for U.S. Energy Policy. Editorial, August 18, 2008.

Foss, M. M., and G. Gulen, 2009. *The Proposed American Clean Energy and Security Act of 2009 and Related Energy/Environment Federal Legislation: Considerations for the Texas Economy*. Final Report to Comptroller of Public Accounts. Center for Energy Economics, Bureau of Economic Geology, Jackson School of Geoscioences, University of Texas, Austin.

GAO, 2008. *Climate Change: Expert Opinion on the Economics of Policy Options to Address Climate Change*. U.S. Government Accountability Office, Washington, D.C.

Gayer, Ted, 2009. *On the Merits of a Carbon Tax*. Testimony before the U.S. Senate Committee on Energy and Natural Resources, December 2, 2009.

Hall, D., 2007. Offsets: Incentivizing Reductions while Managing Uncertainty and Ensuring Environmental Integrity. In: *Assessing U.S. Climate Change Policy Options*. R. Kopp and W. Pizer (eds.). Resources for the Future, Washington, D.C.

Heal, G., and B. Kriström, 2002. Uncertainty and Climate Change. *Environmental and Resource Economics* 22:3–39.

Heritage Center for Data Analysis, 2009. The Economic Consequences of Waxman-Markey: An Analysis of the American Clean Energy and Security Act of 2009. Heritage Foundation, Washington, D.C.

Howarth, R. B., and P. A. Monahan, 1996. Economics, Ethics, and Climate Policy: Framing the Debate. *Global and Planetary Change* 11:187–1999.

ICF International, 2007. *Impact Assessment of Mandatory GHG Control Legislation on the Refining and Upstream Segments of the U.S. Petroleum Industry*, Vol. 1 Report, December 2007. ICF International, Fairfax, Va.

ICF International, 2008. *Impact Assessment of Mandatory GHG Control Legislation on the Refining and Upstream Segments of the U.S. Petroleum Industry, Addendum*. ICF International, Fairfax, Va.

IPCC, 2001. *Climate Change 2001: Impacts, Adaptation, and Vulnerability*. Contribution of Working Group 2 to the Third Assessment Report of the Intergovernmental Panel on Climate Change. Cambridge University Press, Cambridge, U.K., and New York.

IPCC, 2007a. Summary for Policymakers. In: *Climate Change 2007: The Physical Science Basis*. Contribution of Working Group 1 to the Fourth Assessment Report of the Intergovernmental Panel on Climate Change. Cambridge University Press, Cambridge, U.K., and New York.

IPCC, 2007b. Summary for Policymakers. In: *Climate Change 2007: Impacts, Adaptation, and Vulnerability*. Contribution of Working Group 2 to the Fourth

Assessment Report of the Intergovernmental Panel on Climate Change. Cambridge University Press, Cambridge, U.K., and New York.

IPCC, 2007c. Summary for Policymakers. In: *Climate Change 2007: Mitigation of Climate Change*. Contribution of Working Group 3 to the Fourth Assessment Report of the Intergovernmental Panel on Climate Change. Cambridge University Press, Cambridge, U.K., and New York.

Jorgenson, D. W., R. J. Goettle, B. H. Hurd, and J. B. Smith, 2004. *U.S. Market Consequences of Global Climate Change*. Prepared for the Pew Center on Global Climate Change. http://www.pewclimate.org/global-warming-in-depth/all _eports/marketeconsequences/index.cfm.

Kopp, R. J., and W. A. Pizer, 2007. *Assessing U.S. Climate Policy Options*. Resources for the Future, Washington, D.C.

LaCommare, K. H., and J. H. Eto, 2004. *Understanding the Cost of Power Interruptions to U.S. Electricity Consumers*. Energy Analysis Department, Ernest Orlando Lawrence Berkeley National Laboratory, University of California, Berkeley.

Lasky, Mark, 2003. *The Economic Costs of Reducing Emissions of Greenhouse Gases: A Survey of Economic Models*. CBO Technical Paper 2003–3. Congressional Budget Office, Washington, D.C.

Levitan and Associates, 2005. *Post Katrina and Rita Outlook on Fuel Supply Adequacy and Bulk Power Security in New England*. Levitan and Associates, Inc., Boston, Mass.

Linder, K. P., M. J. Gibbs, and M. R. Inglis, 1987. *Potential Impacts of Climate Change on Electric Utilities*. Report 88–2. New York State Energy Research and Development Authority, ICF Inc., Washington, D.C.

Litz, F. T., 2008. *Toward a Constructive Dialogue on Federal and State Roles in U.S. Climate Change Policy*. White Paper, Pew Center on Global Climate Change, Arlington, Va.

Lomborg, B., 2010. Obama Gets Reasonable on the Environment. *Wall Street Journal*, April 17, 2010.

Lorenz, P., D. Pinner, and T. Seitz, 2008. The Economics of Solar Power. *McKinsey Quarterly* June 2008.

Miller, N. L., K. Hayhoe, J. Jin, and M. Auffhammer, 2008. Climate, Extreme Heat, and Electricity Demand in California. *Journal of the American Meteorological Society* 4(6).

MIT, 2007. *Assessment of U.S. Cap and Trade Proposals*. Massachusetts Institute of Technology, Cambridge, Mass.

MIT, 2008. *Appendix D: Analysis of the Cap and Trade Features of the Lieberman-Warner Climate Security Act (S. 2191)*. Massachusetts Institute of Technology, Cambridge, Mass.

MIT, 2009. *Appendix C: Analysis of the Waxman-Markey American Clean Energy and Security Act of 2009 (H.R. 2454)*. Massachusetts Institute of Technology, Cambridge, Mass.

Moulton, D. W., T. E. Dahl, and D. M. Dall, 1997. *Texas Coastal Wetlands*. U.S. Department of the Interior, Fish and Wildlife Service, Southwestern Region, Albuquerque, N.M.

Murray, B. C., and M. T. Ross, 2007. *The Lieberman/Warner American Climate Security Act: A Preliminary Assessment of Potential Economic Impacts.* Nicholas Institute, Duke University, and RTI International, Durham, N.C.

National Assessment Synthesis Team, 2000. Chapter 16 in: *Climate Change Impacts on the United States: The Potential Consequences of Climate Variability and Change.* U.S. Global Change Research Program, Washington, D.C.

Newell, R., and W. Pizer, 2001. *Discounting the Benefits of Climate Change Mitigation: How Much Do Uncertain Rates Increase Valuations?* Pew Center on Global Climate Change, Arlington, Va.

Newell, R., and R. N. Stavins, 2003. Cost Heterogeneity and the Potential Savings from Market-Based Policies. *Journal of Regulatory Economics* 23(1):43–59.

Nordhaus, W. D., 1994. *Managing the Global Commons: The Economics of Climate Change.* MIT Press, Cambridge, Mass.

Nordhaus, W., 2007. *The Challenge of Global Warming: Economic Models and Environmental Policy.* http://nordhaus.econ.yale.edu/dice_mss_091107_public.pdf.

Nordhaus, W. D., and J. Boyer, 2000. *Warming the World: Economic Models of Global Warming.* MIT Press, Cambridge, Mass.

OECD, 2005. *Projected Costs of Generating Electricity 2005 Update.* Organization for Economic Cooperation and Development, Paris.

Parker, L., and B. D. Yacobucci, 2009. *Climate Change: Costs and Benefits of the Cap-and-Trade Provisions of H.R. 2454.* Congressional Research Service, Washington, D.C.

Parry, I. W. H., 2004. Are Emission Permits Regressive? *Journal of Environmental Economics and Management* 47:364–87.

Parry, W. H., and W. A. Pizer, 2007. *Emissions Trading versus CO_2 Taxes versus Standards.* Issue Brief 5. Resources for the Future, Washington, D.C.

Peace, J., and J. Weyant, 2008. *Insights Not Numbers: The Appropriate Use of Economic Models.* White Paper, Pew Center on Global Climate Change, Arlington, Va.

Pew Center on Global Climate Change, 2009a. State Action. *Climate Change 101: Understanding and Responding to Global Climate Change.* Pew Center on Global Climate Change and Pew Center on the States, Arlington, Va.

Pew Center on Global Climate Change, 2009b. *ACELA Summary and Commmparison to the ACES Act.* Pew Center on Global Climate Change, Arlington, Va.

Pew Center on Global Climate Change, 2009c. *Cost of the American Clean Energy and Security Act of 2009 Found to Be Small According to Government Analysis.* Pew Center on Global Climate Change, Arlington, Va.

Pew Center on Global Climate Change, 2010. *A Comparison of the Clean Energy Jobs and American Power Act (CEJAP ACT) and the Carbon Limits and Energy for America's Renewal Act (CLEAR Act).* Pew Center on Global Climate Change, Arlington, Va.

Pielke, R. A. Jr., and C. Landsea, 1998. Normalized Hurricane Damages in the United States: 1925–1995. *Weather and Forecasting* 13:621–631.

Pielke, R. A. Jr., and R. A. Pielke Sr, 1997. *Hurricanes: Their Nature and Impacts on Society.* John Wiley & Sons, Chichester, West Sussex, U.K.

Pigou, A. C., 1912. *Wealth and Welfare*. Macmillan, London.

Regional Greenhouse Gas Initiative, 2003. Memorandum of Understanding. http://www.rggi.org/docs/mou_12_20_05.pdf.

Resources for the Future, 2010. *Summary of Notable Market-Based Climate Change Bills Introduced in the 111th Congress*. Resources for the Future, Washington, D.C.

RMS, 2005. *Hurricane Katrina: Profile of a Super Cat. Lessons and Implications for Catastrophe Risk Management*. Risk Management Solutions, Newark, Calif.

Science Applications International, 2009. *Analysis of the Waxman-Markey Bill "The American Clean Energy and Security Act of 2009."* Report for the American Council for Capital Formation and the National Association of Manufacturers. SAIC, McLean, Va.

Sheraga, J., and F. Sussman, 1998. Discounting and Environmental Management. In: *International Yearbook of Environmental and Resource Economics*. H. Folmer and T. Tietenberg (eds.). Edward Elgar, Cheltenham, U.K.

Sjardin, M., 2008. Cap and Trade Systems: The European Experience. Powerpoint slide presentation. New Carbon Finance, New York.

Smith, J. B., and D. A. Tirpak, 1990. *The Potential Effects of Global Climate Change in the United States*. Hemisphere, New York.

Stavins, R. N., 2007. A U.S. Cap and Trade System to Address Global Climate Change. Hamilton Project Discussion Paper 2007:13. Brookings Institution, Washington, D.C.

Texas Comptroller, 2008. *The Energy Report*. Texas Comptroller of Public Accounts, Austin.

Texas Comptroller, 2009. *2009–2010 Economic Forecast,* Texas Nominal Gross State Product Detail, Fiscal Years 1991–2039. Texas Comptroller of Public Accounts, Austin. http://www.texasahead.org/economy/forecasts/fcsto910/ngsp-fiscal.html.

Texas General Land Office, 2005. *Coastal Texas 2020: A Clear Vision for the Texas Coast*. Texas General Land Office, Austin.

Texas State Data Center, 2009. *Projections of the Population of Texas and Counties in Texas by Age, Sex, and Race/Ethnicity for 2000–2040*. Office of State Demographer, Institute for Demographic and Socioeconomic Research, College of Public Policy, University of Texas, San Antonio. http://txsdc.utsa.edu.

Texas State Data Center, 2010. *Estimates of the Total Populations of Counties and Places in Texas for July 1, 2008, and January 1, 2009*. Office of State Demographer, Institute for Demographic and Socioeconomic Research, College of Public Policy, University of Texas, San Antonio. http://txsdc.utsa.edu.

Weyant, J., F. C. de la Chesnaye, and G. J. Blanford, 2006. Overview of EMF-21: Multigas Mitigation and Climate Policy. *Energy Journal* (Special Issue):1–32.

Yohe, G., 1992. Carbon Emissions Taxes: Their Comparative Advantage under Uncertainty. *Annual Review of Energy* 17:301–26.

CHAPTER 10

Policy

Jurgen Schmandt

The preceding chapters document how climate change and climate variability will impact Texas: temperatures will rise, heat waves will occur more frequently, it will be drier west of the Interstate 35 corridor, severe weather will become more frequent, in-stream flows will fall, biodiversity will decline, and the sea level will rise. The exact timing of these changes and the speed at which they will occur remain uncertain. It is also unknown whether some of the predicted changes will occur gradually or suddenly after a tipping point has been reached.

These findings echo what we presented in the first edition of this book, published in 1995. Results of more recent studies on climate change and Texas include:

- In 1997 the Environmental Protection Agency (EPA) released a report, *Climate Change and the States*, that came to similar conclusions (EPA 1997). Its findings are summarized in Table 10.1.
- A national assessment published in 2000, "The Potential Consequences of Climate Variability and Change," added important points: The summer heat index (which combines temperature and humidity) will increase significantly, heat

TABLE 10.1. Impacts of climate change on Texas, EPA estimates for 2100

CLIMATE CHARACTERISTIC	CHANGE
Temperature	+3.5°F
Precipitation	−5 to −30% in winter; +10% in other seasons
Heat-related illnesses	Increase
Ground-level ozone causing more respiratory diseases	Increase
Flooding	Increase in frequency
In-stream flows	−35%
Sea-level rise	31 inches at Galveston
Harmful algal blooms	Increase

Source: EPA 1997

stress for people and livestock will be more severe, soil moisture will decline as precipitation decreases and evaporation
increases, reductions in water supply and quality will pose
problems primarily for urban and poor populations, the
coastal zone will suffer significant loss of property and damage to ecosystems as a result of coastal flooding and erosion,
oil refineries and the Gulf Intracoastal Waterway will be at
risk from more frequent and more intense storms, and the
health of urban populations will be impaired by an increase in
smog-forming gases from fossil fuel power plants (U.S. Global
Change Research Program 2000).

- Norwine and John (2007) concluded that South Texas by
2100 will be drier, hotter, and stormier. Barrier islands will
have been lost, and saltwater intrusion will diminish water
supplies.

- A recent study by the U.S. Global Change Research Program
emphasizes the risk of more intense droughts: "The consensus
of most climate-model projections is for a reduction of cool-
season precipitation across the U.S. Southwest and northwest
Mexico" (NOAA 2008).

The likely impacts for Texas fall within the range of changes predicted
for North America by Working Group 2 of the 2007 IPCC Assessment
(IPCC 2007): IPCC assigns "very high confidence" to stress and damage
from extreme weather and rising sea level, as well as infrastructure, health,
and safety issues (Table 10.2). Taken together, the findings about the
expected impacts of climate change on Texas have not changed fundamentally over the last decade and a half, but the evidence is now more extensive
and detailed. In contrast, the context for policy development has changed
substantially. There are several reasons for this.

First, decision makers have fewer excuses to defer action because of
scientific uncertainty. Key questions that were controversial two decades
ago have been resolved: Yes, warming occurs not only at the earth's surface
but also at higher altitudes. Yes, increased water vapor in the atmosphere
amplifies global warming. Only the regional distribution of precipitation
and the exact timing of predicted changes remain uncertain and require
more study. Second, climate change is no longer a distant possibility but
is occurring now. "Observational evidence from all continents and most
oceans shows that many natural systems are being affected by regional climate changes, particularly temperature increases." This is in sharp contrast

TABLE 10.2. Climate change impacts on North America

PARAMETER	IMPACT	CONFIDENCE LEVEL
Adaptive capacity	Overly focused on coping with rather than preventing problems	Very high
Extreme weather	Will increase and cause significant economic damage	Very high
Other stresses	Will increase as a result of climate change (infrastructure, health, safety)	Very high
Coastal communities and habitats	Will be stressed in conjunction with development	Very high
Health	Increased risk (heat-wave deaths, water-borne diseases, poor water quality)	Very high
Water	Reduced supply due to diminishing snow pack, higher evaporation	High
Ecosystems	Increased risk of wildfire and insect infestations	High

Source: IPCC 2007:55–56

to our 1995 assumption that the impacts of climate change would be felt only by 2030. Third, the overwhelming majority of scientists are confident that observed and predicted changes in natural as well as social systems are caused by human actions. (These three statements are based on IPCC 2007.) As a result, the urgency to act on climate change has increased greatly since 1995. The 2007–2008 U.N. Human Development report starkly makes the point: "Climate change is the defining human development challenge of the 21st Century. . . . Looking to the future, no country—however wealthy or powerful—will be immune to the impact of global warming" (United Nations Development Program 2007).

In this chapter I briefly review policy development in response to climate change at international, federal, state, and local levels and then discuss policy options for Texas. Several of the preceding chapters also discuss policy issues related to their subject matter. Here I focus on concrete policy measures that should be taken by state agencies and on development of a comprehensive state climate policy.

CLIMATE CHANGE POLICY TO DATE
International and Federal Policies

In 1988, the United Nations convened the Intergovernmental Panel on Climate Change (IPCC) to provide policymakers and the public with reliable summaries of scientific information. The IPCC has since published four assessments of steadily more detailed scientific knowledge about climate change. As mentioned by Bill Dawson in the Introduction to this book, scientists, but not the media, dismiss recent charges of bias raised against the

IPCC. Nevertheless, mistakes were made, and review procedures for the next report, due in 2012, should be tightened.

Toward the end of the first Bush administration, at the environmental summit in Rio de Janeiro, the United States signed the U.N. Framework Convention on Climate Change. The convention marked the beginning of international action against climate change. It soon became obvious, however, that the voluntary measures agreed on in Rio yielded little in the way of results, and a binding agreement was needed.

The Clinton administration took the lead in international negotiations aimed at such an agreement. Against initial opposition from Europe, the United States insisted on a treaty that included provisions for market-based carbon trading. The United States modeled its proposal on the successful sulfur dioxide trading system that had been introduced in the United States as part of the Clean Air Act amendments of 1990 (Environmental Defense Fund 2008; Litz 2008). The resulting agreement, the Kyoto Protocol, was signed by the United States in 1997. The U.S. Senate later passed a resolution (95–0) signaling that it would not ratify the treaty in its original form. Senator John Kerry, an ardent supporter of carbon dioxide regulation, spearheaded the congressional "no" vote, in the belief that it would motivate the president to renegotiate the treaty and make Brazil, China, and India active participants in the fight against global warming. Kerry was confident that a revised treaty would then be ratified (Kerry 2008). He was mistaken. The administration initiated no new negotiations, and the Kyoto Protocol became international law with only two industrialized countries abstaining—Australia and the United States. Australia, after a change in government, joined the treaty in 2008.

As a signatory of the United Nations Framework Convention on Climate Change, the United States currently participates in negotiations (the so-called Bali process) aimed at reaching international agreement on reducing greenhouse gases in the period after 2012, when the Kyoto Protocol will have run its course. Preparatory work on the new treaty has focused on defining an emissions reduction formula that assigns "shared but differentiated" responsibilities to industrialized and developing countries. Judith Clarkson, in Chapter 8, relates details about the Copenhagen climate change summit. Most see it as a failure; however, as a result of President Obama's intervention at the summit, the Big Three (Brazil, China, and India) are now actively engaged in developing a new international treaty.

In April 2007, in response to lawsuits filed by 11 states, three cities, and 13 environmental groups, the U.S. Supreme Court ruled 5–4 that the Clean

Air Act gave the Environmental Protection Agency the authority to regulate carbon dioxide emissions, if the agency found them to be harmful to "public health and welfare." The EPA made a draft finding to that effect but was ordered by the White House not to start regulatory proceedings. The EPA then reversed course and stated that the Clean Air Act was ill-suited for regulating carbon dioxide. Instead, a new law was needed for the purpose (*Washington Post*, July 12, 2008). President George W. Bush, during his second presidential campaign, had promised action to curb carbon dioxide emissions. As it turned out, his administration limited its actions to support of research and voluntary measures.

During the Bush years there was considerable action on climate change in both houses of Congress. In the 107th Congress (2001–2002), four bills setting mandatory limits on carbon dioxide emissions from power plants were introduced: Senate bills S. 556 (Jeffords) and S. 1131 (Leahy) and House bills H.R. 1256 (Waxman) and H.R. 1335 (Allen). A number of other bills related to climate change were also introduced, but only these four included prescribed emission limits or absolute emission reduction goals. In 2003, a McCain-Lieberman bill was defeated 55–43 in the Senate. Both of the senators from Texas voted against the bill. In the 110th Congress (2007–2008), 13 bills mandating limits on carbon dioxide emissions from power plants or multiple sectors of the economy were introduced: S. 280 (Lieberman-McCain), S. 309 (Sanders-Boxer), S. 317 (Feinstein), S. 485 (Kerry-Snowe), H.R. 620 (Olver-Gilchrest), S. 1168 (Alexander), S. 1177 (Carper), S. 1201 (Sanders), H.R. 1590 (Waxman), S. 1766 (Bingaman), S. 2191 (Lieberman-Warner), H.R. 4226 (Olver), H.R. 6316 (Doggett). The Senate, in June 2008, defeated the Lieberman-Warner Climate Security Act, which would have introduced a market-based cap-and-trade system for carbon dioxide. In October 2008, Representatives Dingell and Boucher released a discussion draft; it was never introduced as legislation, but its introduction by a sitting committee chair made it significant.

The Obama administration brought a dramatic shift toward developing an aggressive climate change policy, reinvigorating action by Congress as well as the EPA. Again, these measures are summarized in the Introduction and previous chapters. Suffice it to say here that as of this writing (April 2010), Congress has not enacted legislation, EPA regulations are being contested in court, and the public debate ranges from skepticism to lukewarm support.

The Climate Action Partnership, an influential group supporting federal legislation, was founded in 2007. Founding members include a number of major corporations (Alcoa, BP America, Caterpillar, Duke Energy, DuPont, FPL Group, General Electric, PG&E Corporation, and PNM Resources) and four nongovernmental organizations (Environmental Defense Fund, Natural Resources Defense Council, Pew Center on Global Climate Change, and World Resources Institute). These organizations "have come together to call on the federal government to quickly enact strong national legislation to require significant reductions of greenhouse gas emissions." Early in 2009 the partnership published *A Blueprint for Legislative Action,* calling for "a mandatory, national economy-wide climate protection program that includes aggressive emission reduction targets for total U.S. emissions and for capped sectors" (Climate Action Partnership 2009).

States, Regions, and Cities

While waiting for federal action, over the course of the last decade, states and cities have become major players in attempts to reduce greenhouse gases and increase energy security. In the 50-year history of environmental policy, the states have repeatedly spearheaded federal action. This was the case in early action on air and water pollution, as well as acid deposition. The current state initiatives involve more states and spawn more cooperation among them than ever before. Lutsey and Sperling (2008), in a detailed analysis of this trend, write: "Local, regional, and state governments are now following a prescribed pattern of inventorying their emissions, establishing climate change action plans, setting emission reduction targets . . . , enacting state-level regulations and standards explicitly targeting greenhouse gases, and forging multi-government alliances to reinforce and support their actions." Temporarily, at least, environmental policy in the United States has shifted to a decentralized bottom-up approach.

An EPA scorecard of state actions to combat climate change identifies 15 programmatic initiatives that the states have launched between 2000 and August 2008 (Table 10.3). More details are given in Chapter 8, and much of the information is taken from the Pew Center's website on global climate change (Pew 2008a). The research conducted by the Pew Center documents a fast-rising level of state programs, currently involving more than half of all states, acting either jointly with neighboring states or on their own. State laws authorizing these initiatives are also listed on the Pew Center site (Pew 2008b). Information on specific bills has been compiled by the National Caucus of Environmental Legislators (2008). Actions taken by cities and counties are listed on the website of the U.S. Conference of Mayors (2007, 2010). Each of these is discussed below.

TABLE 10.3. State initiatives on climate change

INITIATIVE	NUMBER OF STATES PARTICIPATING, APRIL 2010		TEXAS PARTICIPATION, APRIL 2010
	YES	NO	
State advisory board	28	23	No
Member of regional initiative	30	21	No
Greenhouse gas inventory	30	21	Yes
Climate change action plan	34	17	No
Statewide greenhouse gas target	19	32	No
Statewide greenhouse gas cap	6	45	No
Electricity disclosure	21	30	Yes
Greenhouse gas registry	40	11	No
Mandatory greenhouse gas registry	19	32	No
CO_2 offset requirements	3	48	No
Greenhouse gas performance standard	4	47	No
Advanced coal technology	14	37	Yes
Power sector greenhouse gas cap and trade	10	41	No
Greenhouse gas auto standards	15	36	No
Low carbon fuel standard	1	50	No

Source: Compiled from U.S. Environmental Protection Agency website, http://www.epa.gov/climatechange/wycd
/stateandlocalgov/index/html, accessed April 23, 2010

California

As has been the case in response to other environmental threats, California leads the states in acting on climate change. With bipartisan support from policy leaders and broad public acceptance, California has put in place an ambitious program to reduce emissions. Industry response was divided, but support eventually outweighed opposition. Several factors make it easier to act on climate change in California than elsewhere. Republicans and Democrats, after much wrangling, reached a meaningful compromise. There is no coal or automobile lobby in the state. The state economy, to a considerable extent, is focused on itself and markets across the Pacific. Consumers are more willing to adopt new solutions. Texas can learn useful lessons from several of California's programs.

In 2002, California passed a law to reduce greenhouse gas emissions from automobiles and trucks. New greenhouse gas emission standards for light-duty vehicles were promulgated in 2004. This was followed by the linchpin of California's initiative, the Global Warming Solutions Act of 2006 (AB 32). The bill mandates a reduction in total emissions, from the 1990 base, of 28 percent by 2020 and 80 percent by 2050. These numbers equal the emission reduction targets that have been enacted or proposed by members of the

European Union. Once Governor Schwarzenegger had signed the enabling legislation, implementation was entrusted to state agencies that work closely with university and industry experts. The lead agency is the California Air Resources Board (CARB), a department of the California Environmental Protection Agency. The legislation grants the CARB wide-ranging powers to set policies, draw up regulations, lead the enforcement effort, levy fines and fees to finance it, and punish violators. CARB prides itself on its technical expertise and relative insulation from political interference. Half of the board's 11 members are scientific or professional experts; the other half represent regions of the state. Mary Nichols, the chair of CARB, was assistant administrator for EPA's Air and Radiation program, where she gained experience implementing the sulfur dioxide cap-and-trade system.

CARB began its work with a detailed scoping study, then developed a precise timetable, and is now drafting mandatory rules that set energy efficiency standards for vehicles, appliances, and housing (California Air Resources Board 2008a). CARB claims that the measures needed to reach the 2020 reduction goal, despite high costs, will benefit the state economy and job market. "Under the Plan, homeowners can achieve electricity savings between 1,500 and 1,800 kWh per year for older and newer homes, respectively, and over 300 therms of natural gas per year" (California Air Resources Board 2008b). These savings would continue California's success in holding per capita electricity use constant since 1970, while per capita use rose by nearly 80 percent in the United States as a whole.

The main goal of the 2006 act is to reduce emissions from the transportation sector (which accounts for 39 percent of emissions), but all sectors of the economy, as well as private residences, must reduce emissions. Electricity generation currently accounts for 28 percent of emissions; residences contribute 9 percent and refineries 8 percent.

The climate change act is to take effect in 2012. As a result of rising unemployment in the state, it has lost some public support. Two Texas oil firms with operations in California are supporting a ballot initiative to halt implementation of the act until unemployment in California falls from its current 12.5 percent level to 5.5 percent (*New York Times*, April 7, 2010).

The 2004 vehicle standards have been opposed by the automobile industry, and the EPA in 2008 denied the necessary waiver for California to proceed with the new standard. One of Barack Obama's first acts as president, on January 26, 2009, was to direct the EPA to reconsider the case. On June 30, 2009, the EPA approved California's greenhouse gas emission standards. The California emission standard is now the basis for federal standards issued by the EPA and the Transportation Department in March 2010. This will finally lead to a nationwide standard that is expected to cut emissions of carbon dioxide and other heat-trapping gases by about 30 percent from 2012 to 2016.

California expects short-term emission reductions to result from new low-carbon fuel standards. As the downside of the grain-based ethanol mandate becomes clearer, this goal may be difficult to meet. Long-term improvements will result from technical innovation. For this the state relies on a clear division of labor: the state issues performance standards and monitors compliance, and industry will develop successful solutions. It is not the role of state government to pick winning technologies or industries. An education campaign is under way to encourage people to drive less. Cities and counties are required to develop their own carbon reduction strategies (Sperling 2008).

CARB is also engaged internationally, sharing engineering and policy expertise on the measurement and control of greenhouse gases with regions such as Brazil's Amazon states and Indonesia's forested provinces.

Regional Alliances

Thirty states have joined regional climate change initiatives. The first such alliance, initiated in 2005, was the Northeast Regional Greenhouse Gas Initiative; it was followed in 2007 by three others (Fig. 10.1, Table 10.4). Twenty of the participating states have Democratic governors; the others have Republican governors. Among the participating states, 15 legislatures are controlled by Democrats, 6 are controlled by Republicans, and 9 are split between House and Senate. The alliances, therefore, have attracted significant bipartisan support. It is noteworthy that the entire Southeast region of the country and Texas are not participating. Florida decided to develop its own cap-and-trade emission reduction program (Florida Energy and Climate Change Action Plan 2008).

The Northeast alliance, now called the Regional Greenhouse Gas Initiative, Inc. (RGGI), conducts quarterly auctions for carbon dioxide allowances for fossil fuel-based power plants in the Northeast. The first auction took place in September 2008. The spring auction in 2010 yielded $88 million for investment in clean energy technologies (RGGI 2010). It seems that RGGI avoided the initial mistakes made by the European cap-and-trade system that became operational in 2005. During its first phase, the European program suffered from incomplete data, wrong assumptions, and overallocation of carbon dioxide allowances. The program is now in its second phase and considered to be a success (Sjardin 2008).

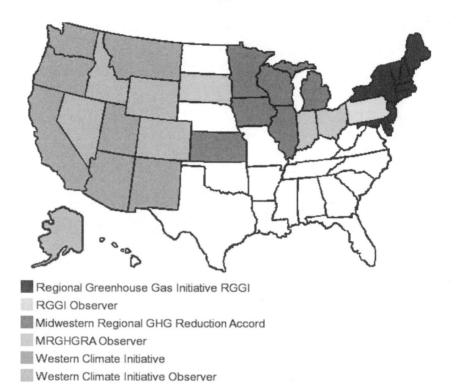

Regional Greenhouse Gas Initiative RGGI
RGGI Observer
Midwestern Regional GHG Reduction Accord
MRGHGRA Observer
Western Climate Initiative
Western Climate Initiative Observer

FIGURE 10.1. Regional climate change initiatives in the United States. *Source:* Pew Center on Global Climate Change.

Cities

In 2005 a bipartisan coalition of 132 U.S. mayors, led by Seattle mayor Greg Nickels (D) and New York City mayor Michael Bloomberg (R), committed to reducing their municipal greenhouse gas emissions to 7 percent below 1990 levels by 2012, in line with the Kyoto treaty targets. This is to be achieved by a wide range of measures, including urban reforestation, building standards that increase energy efficiency, changes in landfill practices, and public education campaigns. The mayors refuted the Bush administration's argument that the Kyoto Protocol would devastate the economy. Instead, the mayors stated that they signed on precisely for economic reasons. Nickels was jarred by a series of dry winters that threatened Seattle's drinking water and hydropower sources. The mayor of Bellevue, Nebraska, was worried about the effects of droughts on farms. The mayor of New Orleans was concerned about the effects of rising sea levels on "the very existence of New Orleans" (Sanders 2005). By 2010, 1,017 cities had joined the coalition (U.S. Conference of Mayors 2010).

TABLE 10.4. Regional climate change alliances

	MEMBERS			
ALLIANCE	FULL	OBSERVER	FOREIGN	PROGRAM
Northeast Regional Greenhouse Gas Initiative	10[a]			Implement mandatory cap-and-trade market for CO_2 emissions from fossil fuel plants
Energy Security and Climate Stewardship Platform for the Midwest	11[b]			Increase energy efficiency, renewable energy sources, and biofuel production. By 2010 develop regulatory framework for capture and storage of carbon. By 2012 site and permit regional CO_2 transport pipeline. By 2020, new coal power plants will capture and store CO_2 emissions
Midwestern Regional Greenhouse Gas Reduction Accord	6[c]	4	4	Reduce greenhouse gas emissions 60–80%. Develop greenhouse gas reduction tracking system. Adopt low-carbon fuel standards. Develop cap-and-trade market for CO_2 (observers will participate)
Western Climate Initiative	7[d]	6	7	Develop regional emission target. Establish market-based system for CO_2 control. By 2020 reduce greenhouse gas emissions 15% below 2005 levels

[a] CT, DE, MA, MD, ME, NH, NJ, NY, RI, VT

[b] IA, IL, IN, KS, MI, MN, MS, ND, NE, OH, WS

[c] IL, IA, KS, MI, MN, WS

[d] AZ, CA, MT, NM, OR, WA, UT

Source: Pew Center on Global Climate Change and websites of the regional climate change alliances

A survey of member cities conducted by the U.S. Conference of Mayors showed that of 134 cities reporting, 80 percent use renewable energy; 97 percent have switched to energy-efficient lighting in public buildings, street lights, and traffic signals; 72 percent power city vehicles with alternative fuels; 90 percent require new city buildings to be energy efficient and environmentally sustainable; and 70 percent encourage private contractors to build energy-efficient structures. Almost all cities consider their actions on global warming to be direct contributions to improved health and quality of life for their citizens (U.S. Conference of Mayors 2007).

How significant are state and local actions? Lutsey and Sperling (2008) calculate that combined state and local emission reduction plans apply to 43 percent of U.S. greenhouse gas emissions. "If the 17 states that have set their own GHG emission-reduction targets (generally to 1990 levels by the year 2020) in fact were to achieve those targets, nationwide U.S. GHG emissions would be stabilized at 2010 levels by 2020—without any serious mitigation action taken by over half of the states. . . . Although these reductions are nowhere near the deeper longer-term reduction that would be required for climate stabilization, they are nonetheless substantial and significant relative to federal inaction." As regional and local programs mature, states and cities increasingly learn from each other and use common protocols and standards for their actions. Several years back it looked as if an uncoordinated patchwork of local and regional rules was emerging. By now, the sum of state and local actions has been described as a "consistent U.S. policy structure" based on coordinated inventories of emissions, mitigation plans, and emission reduction targets.

This may be an overly optimistic conclusion. The pros and cons of bottom-up policies in the context of environmental federalism need to be clearly recognized. On the negative side, a patchwork of state or regional regulations imposes cumbersome burdens on industry, wastes regulatory resources, cannot deal efficiently with cross-boundary pollution, encourages polluters to move to more lenient jurisdictions, and imposes costs on regulating states without guarantees that they will reap proportionate benefits. There are also advantages: state initiatives can help the federal government re-engage in the development of domestic and international policy; the experience gained by the states can guide federal policy development and define U.S. policy preferences. Above all, local and regional involvement is essential to achieve the technological, economic, and social transformations that large reductions in carbon dioxide will require. Strong local commitment is a prerequisite for successful federal regulation (Lutsey and Sperling 2008).

It is unclear whether federal legislation, if and when it is enacted, will supersede regional programs or accommodate them. I share the view of the Pew Center on Global Climate Change: "Federal action on climate change is needed to achieve the significant reductions science demands and to establish a minimum level of uniformity across the U.S. economy. This federal action can preserve room for states to continue in their important roles as policy innovators, on-the-ground implementers, and policy drivers, and to capitalize on the significant experience in the states across the many aspects of climate change" (Litz 2008).

RESPONDING TO CLIMATE CHANGE IN TEXAS

In the 1995 edition of this book, we recommended that Texas develop policy in response to three threats from climate change:

1. *Reduction of carbon emissions.* Texas is home to 25 percent of the nation's refining capacity and leads the country in

emitting greenhouse gases. We argued that the state would be well advised to contribute proactively to the national debate about the costs and benefits of a federal carbon tax compared to a cap-and-trade system. State policymakers needed to know how either measure would impact the Texas economy. Would the state be hurt more than other states if the new charges were imposed at the refinery, which would be the case if a cap-and-trade system is introduced? Or would it be better to have the charge paid by the consumer, which would be the case if a carbon tax is imposed?

2. *Sea-level rise.* The gentle slope of the Texas coastline will expose large coastal areas to rising sea levels, erosion, and saltwater intrusion into aquifers. The state should use its experience with land subsidence from pumping oil, gas, and water to incorporate sea-level rise into its coastal management plan, as well as zoning ordinances and insurance requirements.

3. *Water scarcity.* A large part of Texas is semiarid. The Panhandle, West Texas, and the Rio Grande border region, in particular, will experience higher temperatures, higher rates of evaporation, and decreased rainfall. Combined with population growth, this will lead to greater demand for water and increased competition among agricultural, municipal, and ecological water demands. State water management agencies should prepare for global warming through conservation, more efficient water use, and better drought preparedness.

Actions to Date

Texas did not follow these recommendations. As a matter of fact, global warming rarely appeared on the radar screen of state policymakers after the initial wave of interest in the early 1990s had dissipated. A scorecard prepared by the Environmental Protection Agency shows that Texas acted on only 3 out of 15 climate change initiatives that are currently under way in various states (Table 10.3):

1. Texas completed a greenhouse gas inventory in 2002. As a follow-up step, the EPA recommended that the state regularly report its greenhouse gas emissions. This was not done.

2. As of July 2002, retail electric providers must provide the standardized-format Electricity Facts Label to customers when

they request it. Labels must include electricity prices, contract terms, and sources of generation with emissions levels.

3. The 2005 Legislature passed H.B. 2201, providing for the transfer of carbon dioxide from a future clean coal power plant to the Railroad Commission (for injection into old oil and gas wells) and the Texas Water Development Board (for injection into old water wells). Legislators also approved expedited permitting for such projects. The bill was intended to help Texas win the national competition for a clean coal demonstration plant.

In contrast, there has been considerable action on the part of local governments. Twenty-three Texas cities have joined the Mayors' Climate Protection Agreement discussed in the previous section. Signatories include Austin, College Station, Dallas, El Paso, Fort Worth, Laredo, and San Antonio.

The City of Austin has set itself the goal of becoming "the leading city in the nation in the fight against global warming" (City of Austin Web 2007). The city council adopted a broad-ranging climate protection plan in 2007. The plan includes action on several fronts. All city vehicles, facilities, and operations will be carbon-neutral by 2020; the city's electric utility will reduce emissions through large increases in conservation, efficiency, and renewable energy sources; existing utility plants will be retired early and replaced by carbon-neutral plants; building codes for residential and commercial properties will require high energy efficiency; a community plan is being developed with provisions for reducing greenhouse gas emissions from all sources (City of Austin Website 2007). The city-owned utility company has been instructed to get 30 percent of its power from renewable sources by 2020. The city council, on August 28, 2008, approved a 20-year $2.3 billion contract with a private provider to build and operate a 100-megawatt biomass plant that will use wood waste as fuel (*Austin American-Statesman,* August 29, 2008). On Earth Day 2010, the city council approved an energy efficiency plan that will increase Austin's renewable energy use to 35 percent, up from the current 12 percent (*Austin American-Statesman,* April 12, 2010). To report on progress and solicit citizen input, a special climate-change website went online in June 2008 (City of Austin Web 2008). The city council approved recommendations by a special task force on energy efficiency requirements for new buildings (Austin 2007). These requirements have since been incorporated into the city building code. A second task force proposed that homes older than 10 years must have a mandatory energy audit at the time of sale. Commercial and multifamily properties would have an audit performed within 2 years. Upgrades will be voluntary and performance will be monitored every 2 years. The city council approved these recommendations on June 1, 2008.

There has also been action on the part of Texas-based foundations. In February 2008, the George and Cynthia Mitchell Foundation created a three-year $6 million program to advance renewable energy and energy efficiency technologies and to help Texas develop a comprehensive global warming strategy (Mitchell Foundation 2008). The Houston Endowment has made several grants to Texas universities and environmental groups in support of research and education on global warming, air quality, and energy efficiency. One of their grants has made possible the publication of this book. The Houston Advanced Research Center has used grant money to publish "Texas Climate News: Reporting on Climate Change and Sustainability Issues" (http:texasclimatenews.org/Default.aspx).

City and foundation activism may have helped to generate more interest in climate change during the Texas 2007 legislative session. Twelve bills were introduced to curb greenhouse emissions or to support precautionary measures to adapt to global warming. Yet, not much was accomplished. Seven bills never received a hearing. Four died after hearings in committee. Only Senate bill 1762 passed, charging the Texas Water Development Board (TWDB) with studying the likely impact of climate change on drought conditions in Far West Texas. This may trigger the TWDB to change the position it adopted in a 2007 report to the Legislature: "When considering the uncertainties of population and water demand projections, the effect of climate change on the state's water resources over the next 50 years is probably small enough that it is unnecessary to plan for it specifically" (TWDB 2007a:299).

The 2009 Legislature passed Senate Bill 184, calling on the Texas Commission on Environmental Quality to submit a report to the Legislature by the end of 2010 on ways businesses and consumers can reduce greenhouse gas emissions that also reduce costs or at least are financially neutral. Other bills related to climate change failed to pass.

Overall, the majority of state leaders remain skeptical about global warming. Efforts to improve energy efficiency have a better chance to succeed in the Legislature if the bill does not mention climate change. According to a study published in 2000, the Legislature is still at the point that was then the norm for U.S. cities: "Ironically, the most effective way to get municipal governments to take action on global climate change is to not talk about global climate change" (Betsill 2000).

Sector-specific Policy Measures

What should Texas do to respond to climate change? The Pew Center on Global Climate Change, in its "Agenda for Climate Action," urges action in six areas: science and technology research, market-based emissions management, emissions reduction in key sectors, energy production and use, adaptation, and international engagement (Pew 2006). In the following sections, I discuss how Texas might become active in several of these areas, by contributing to the national policy debate, joining the efforts of other states, and acting on its own to prepare for a changed climate.

The time for this to happen seems right, as the mood in Texas is shifting from inaction to support of technological innovation and adaptation strategies. Larry Soward, one of three commissioners appointed by the governor to the Texas Commission on Environmental Quality, advocates a more active strategy: "As the nation's leading emitter of greenhouse gases, and with an extremely vulnerable coastline, it only seems reasonable and logical to me for us here in Texas to step up, take a leadership role, and begin to seriously and meaningfully address our greenhouse gas emissions" (*Austin American-Statesman,* April 30, 2008). In testimony before the Texas Senate Natural Resources Committee on July 8, 2008, Soward proposed, as an immediate first step, legislation "requiring Texas to either develop and implement our own greenhouse gas inventory and registry, or require us to join the multistate Climate Registry." Tony Bennett, writing for the Texas Manufacturers Association, favors long-term investments in innovative technologies: "Policies should encourage technology investment to safeguard the environment without decimating the economy. . . . Misguided and unrealistic mandates force employers to divert resources in the near-term rather than promote spending for long-term innovations to reduce greenhouse gases and increase efficiency" (*Austin American-Statesman,* May 12, 2008).

Yet public support remains weak, as several opinion surveys have shown. "Climate change and climate variability *do not* emerge as top of mind problems" (Vedlitz 2004). "Texans are very concerned about drought. They see climate change as relevant, but less important than other factors" (Vedlitz 2008). Not surprisingly, energy efficiency fares better. A 2009 survey by the George and Cynthia Mitchell Foundation found that Texans are willing to pay for investments in alternative energy (Baker 2009). The governor remains staunchly opposed to federal regulation of greenhouse gases that would, in his opinion, "run this nation's strongest economy right off the tracks and into the ditch" (*Austin American-Statesman,* November 26, 2008).

The rest of this chapter presents policy recommendations in the areas of science policy, emissions control, water management, and coastal management, concluding with an outline of a comprehensive Texas policy in response to climate change.

Science Policy: Make Results Relevant to Resource Managers

Managers of air sheds, water basins, and ecological regions have found it difficult to integrate scientific information provided by climatologists into their plans and actions. Hydrologists, for example, use the catchment area as the basis for planning. They need data on precipitation, runoff, in-stream flow, evapotranspiration, and related factors to estimate future water supply. Climatologists, by contrast, use square grids of equal size to model changes in temperature and precipitation. To date, climatologists and hydrologists do not speak the same language. To make climate model results more compatible with the needs of water managers (and other managers of natural resources), new assessment methodology and closer interaction with practitioners in the field are needed. The problem is significant but not sufficiently recognized. It warrants detailed discussion.

When climatologists study global warming, they ask how an increase in the concentration of greenhouse gases in the atmosphere is likely to change temperature, precipitation, sea level, ocean currents, and other factors that determine the climate of the globe. Using increasingly sophisticated climate models, climatologists provide reliable information about expected temperature changes at different latitudes. This makes it possible to predict temperature changes at regional scales. For a long time, however, the square grids used by the global circulation models were too large to be of much use to regional resource managers. By now, the models have been refined to produce information for horizontal grids measuring 200 square kilometers, about 77 square miles, and some regional models have been scaled down to 100 square kilometers, occasionally even 10 square kilometers. Even so, certain problems with down scaling remain unresolved, in particular when predicting changes in precipitation. There are several reasons for this. First, different models still give different results regarding where to expect more or less rain. Second, square grids do not easily translate into the planning units that are used in water, air, and ecosystem management, such as water basins, air sheds, and ecosystems. Third, climatologists and resource managers use different data sets. Most important, climate change research focuses on a single problem, namely, the increased presence of greenhouse gases in the atmosphere and the consequences of this phenomenon. How climate change impacts existing problems in the water basin or air shed is not considered. For all these reasons, it remains difficult for resource managers and policy-makers to integrate climate change data into regional planning and action.

The National Academy of Sciences has proposed methodology to over-come these shortcomings. What is needed, in the academy's view, is an integrated understanding of regional problems that, taken together, threaten sustainability. Traditional environmental threats deal with a specific distur-bance of a natural system by human action. Examples include water and air pollution, contamination by toxic substances, depletion of the ozone layer, or climate change. In each case, a single man-nature interaction is analyzed. The complex chain linking cause to effect is identified. Over time, scientists have successfully used the traditional tools of science (formulating a hypoth-esis, observation, measurements, interpretation, and linking data to theory) to explain the issue and provide policymakers with a knowledge base that can guide remedial action.

Single-issue problems, however, in the view of the academy, are no longer the key obstacles standing in the way of sustainable development. People live in specific places where environmental threats arise from problem clusters, some caused by global forces, others due to regional factors. These problem clusters must be understood before reliable policy advice can be given. Many regions suffer from multiple, cumulative, and interactive stresses, driven by a variety of human activities. Here are two examples: some regions experi-ence the combined stresses of population growth, water pollution, and ill health; other areas experience the joint effects of soil depletion, drought, and malnutrition. Climate change is an additional problem that needs to be integrated into the existing set of regional problems. Problem clusters of this kind are difficult to unravel and complex to manage. They are shaped by physical, ecological, and social interactions in particular places. *They are place-based.* An academy report recommends developing sustainability science as a new approach that can unravel complex problem clusters into their constituent components, follow their interactions, present an integrated view of the issue, and identify options for workable solutions. The report concludes, "Developing an integrated and place-based understanding of such threats and the options for dealing with them is a central challenge for . . . a transition towards sustainability" (National Research Council 1999:8). Ralph Cicerone, president of the Academy of Sciences, offers this definition: "Sustainability science is supposed to draw upon contributions from every field of science and engineering and medicine, social science, and draw upon the observations of the direction we're headed and the likely outcomes and continue to navigate this transition to a more sustainable direction" (personal interview, March 15, 2007).

Successful sustainability science is a precondition for successful policy development. Both the science community and the policy community must understand this. Scientists must organize in teams of experts from relevant scientific disciplines who work together to untangle a local or regional problem cluster, identify interactions and feedbacks between contributing causes, and construct an integrated, policy-relevant knowledge base. This is not a revolution in scientific method but a change in research organization. Sustainability science requires scientists to organize their work cooperatively by working in teams whose members are trained in different disciplines. Team members will use the methods they have been trained in to study parts of the problem cluster. They will then work as a group to integrate their findings. They will also take the time to interact closely and repeatedly with stakeholders and decision makers. Finally, they will present their results to the public and decision makers in a format that is accessible to the nonspecialist. Policymakers and resource managers must call for and use this kind of high-level risk and scenario analysis so that their decisions rest on firm ground.

Two Texas scientists have suggested how the concept of user-focused, place-based sustainability science can be implemented and used to address climate change issues in the state. Both proposals emphasize two points. First, climate change needs to be considered in conjunction with other regional problems, not as a stand-alone issue. Second, there must be better links between the research and practitioner communities. Eric Barron, until recently dean of the Jackson School of Geology at the University of Texas in Austin and now director of the National Center for Atmospheric Research, proposes the creation of "environmental intelligence centers" where research on the climate is linked to human activities. Multiple stresses are studied at local and regional scales. The centers will specialize in what Barron calls stage 2 sciences, focusing on the linkage between prediction and action (Barron 2008). Robert Harriss, president of the Houston Advanced Research Center, advocates the creation of "urban sustainability centers" to "catalyze, facilitate, and support the integration process necessary to creating use-inspired solutions to . . . a renewable energy future, adaptation to climate change, sustaining biodiversity and ecosystem services, reducing vulnerabilities to pollution and natural disasters" (Harriss 2008). These proposals aim at marshaling science to the solution of today's problems in the same way that the agricultural research and extension services have contributed to the success of agriculture in America and Texas for the last century and a half.

Action is needed on three fronts. Texas universities and research organizations should encourage and reward stage 2 science projects that are interdisciplinary, problem-focused, and participatory. The Texas Agricultural Extension Service, or some other interested state agency, should prepare a concept paper and action plan for the creation of urban service centers. Texas-based foundations should provide support for these initiatives.

Energy, Environment, and Economy: Reducing Emissions

Up to now Texas has declined invitations to participate in regional efforts to regulate greenhouse gases. Nor have the governor and the Legislature encouraged the Texas Commission on Environmental Quality to regulate carbon dioxide as an air pollutant. Texas's continued failure to participate with others or act on its own may give other states an edge in influencing national policy during its formative stages. Actions by regions and states are getting business and people ready for the low-carbon future.

Texas does, however, invest in programs aimed at improving energy efficiency and independence. Pursuing these strategies benefits the state's economy, creates jobs, and reduces costs over time. Whether intended or not, programs of this kind also reduce greenhouse gas emissions. Geographically, Texas is well placed to lead in the development of solar and wind power technologies. The state can also be a player in the search for better biofuels, in particular from switchgrass, wood chips, and algae from the Gulf of Mexico. As native oil reserves decline, more of the state's energy needs, at least for the next 20 to 30 years, will need to be met by coal and nuclear technology, which currently account for about half of electric power produced in Texas (the other half coming from natural gas). Ten new coal plants have been permitted or are waiting for approval. Three more are on the planning board. Without clean coal technology, the state will vastly increase its carbon dioxide emissions and impede its ability to meet future federal emission reduction mandates. Natural gas offers itself as a less CO_2-intensive alternative. It is now even more attractive, because new technology, developed in the 1990s by Texas-based Mitchell Energy, has made it possible to extract gas economically from large shale formations in North Texas (Potential Gas Committee 2009).

A cost-effective strategy to reduce energy demand can be pursued by improving energy efficiency in homes, businesses, and transportation. Results will be driven by market forces (the cost of energy and changes in consumer behavior) rather than regulation. Experts estimate that 30 percent of energy, and often more, is wasted and could be spared using currently available technologies. "In my team's latest redesigns for $30 billion worth of facilities in 29 sectors, we consistently found about 30 to 60 percent energy savings that could be captured through retrofits, which paid for themselves in two to three years. In new facilities, 40 to 90 percent savings could be gleaned—and with nearly always lower capital cost" (Lovins 2008). Such gains will benefit the economy and immediately lead to significant reductions in greenhouse gas emissions. The state and cities have a dual role in striving to achieve these goals: issuing more stringent standards for buildings, appliances, and vehicles and monitoring compliance.

Development of new technologies should be driven by business investments. The state should not make these decisions. It should lend support in the form of research grants and demonstration contracts. Candidates for support include clean coal technology, carbon dioxide capture and storage, next-generation nuclear power, safe storage of nuclear fuels, and alternative fuels. The value of state leadership on these issues has been demonstrated in California. According to the December 1, 2005, summary by the Hewlett Foundation, state policies encouraging the use of natural gas and renewable resources, as well as the aggressive promotion of conservation measures, have resulted in a drop in per capita emissions of a third since 1975, while the nation's per capita emissions have stayed flat. This has resulted in savings of approximately $1,000 per year in electricity costs for each Californian and has helped the economy grow an additional 3 percent. The job growth created by the energy-efficiency industry will generate an $8 billion payroll over 12 years (Friedman 2008:279).

Some of these initiatives are already under way. The State Energy Conservation Office reports: "The U.S. wind industry grew by 45 percent in 2007, and over half of that growth was contributed by Texas. Texas is the leading wind state in the United States, accounting for close to one third of the nation's total installed wind capacity, which is the equivalent of the electricity needed to power more than one million Texas homes. A single megawatt of wind energy can produce as much energy used by about 230 typical Texas homes in a year" (SECO 2008). The growth of the state's wind power industry has been encouraged by the Legislature, which created the Texas Renewable Energy Portfolio Standard in 1999. Initially the Texas standard mandated that utility companies jointly create 2,000 megawatts of renewables by 2009, based on their market share. In 2005, Senate bill 20 increased the requirement to 5,880 megawatts by 2015, of which 500 megawatts must come from nonwind resources. The bill set a goal of 10,000 megawatts of renewable energy capacity for 2025 (Rabe 2006). This bill also required the Public Utility Commission to design plans for new transmission lines to bring this power from West Texas and the Panhandle to urban areas in the state. In July 2008, the Public Utility Commission of Texas gave preliminary approval to funding and construction of new transmission lines that will bring 18,456 megawatts of power to the eastern half of the state. The project will cost $5 billion, will be completed in four to five years, and will cost the average consumer $4 per month (Environment News Service 2008).

In mid-2008 the installed capacity for wind power in Texas amounted to 5,300 megawatts. At the end of 2009 more than 10,000 megawatts had been installed. Texas is the leader among states, followed by Iowa, California, Washington, and Oregon. An important part of the Texas initiative is a plan promoted by T. Boone Pickens to build the nation's largest wind farm, capable of producing 4,000 megawatts. The project is scheduled to be completed in 2014 but may take longer because of the worldwide economic crisis. The state's commitment to installation of transmission lines is an important step in the expansion of renewable energy production. Another hurdle is the need to match demand and supply over the course of the day. Wind speeds can be highly variable, and the development of energy storage facilities would make wind energy a more dependable source of power. A design study for underground storage of wind energy using compressed air found that excellent geological conditions exist in parts of Texas, New Mexico, and Oklahoma for this technology. The study concludes that development of compressed air energy storage in the study area "could realize approximately $10 million per year in net value [and] integrate an additional 500 MW of wind" (Ridge Energy Storage and Grid Services 2005). The U.S. Department of Energy, in a national competition, selected Austin, Houston, and San Antonio as Solar America Cities, to receive cash grants and technical assistance from National Laboratories.

Taken as a whole, new sources of energy are still in their infancy, partly because the technology is not yet mature, partly because prices remain high. A recent estimate predicts that solar energy in the United States will represent only about 3–6 percent of installed electricity generation capacity, or 1.5–3 percent of output, in 2020 (Lorenz et al. 2008). Geographically, the greatest potential for solar energy is in the American Southwest, but Far West Texas also enjoys sufficient daily radiation to participate in the development of solar power. An ambitious plan by a group of industrialists and scientists proposes a solar-based energy system that integrates photovoltaics, compressed air energy storage, concentrated mirror-derived solar power, and a new DC transmission grid. The authors claim that solar energy by 2050 can provide 69 percent of electricity and 35 percent of total energy. The plan would require federal subsidies over 40 years of more than $400 billion (Zweibel et al. 2008). A more modest and probably more realistic estimate of the future role of alternative fuels in meeting the energy needs of the state is presented in Chapter 9. In April 2008, Governor Perry announced that Texas would invest $1 million through the Texas Enterprise Fund in Heliovolt Corporation of Austin for the construction of a 125,000 square foot manufacturing facility and development space to test and produce the company's thin-film solar power cells. This investment, according to the Governor's Competitiveness Council (2008), will create nearly 160 jobs and $62 million in capital investment.

For several years now there has been intensive debate on testing clean coal technology in some of the new coal-powered plants planned for Texas, but no firm commitments have been made. The 2005 Legislature approved accelerated screening and permitting for clean technology plants. Environmental groups opposed a plan by NRG Texas to build a new coal-fired power plant in Limestone County. In July 2008, the Environmental Defense Fund dropped its opposition in exchange for NRG's offer to reduce carbon dioxide emissions from the new plant by half, using sequestration or investing in CO_2-absorbing plants (*Austin American-Statesman*, August 18, 2008). The Texas proposal to the U.S. Department of Energy for building a clean coal power plant was not successful, but planning work along this line is continuing at the Texas Railroad Commission. The Jackson School of Geology at the University of Texas conducts large studies of carbon dioxide sequestration and storage. Significant research is also under way at the University of Texas on algae as a source of biofuel. The 2009 Legislature passed H.B. 469, which establishes a two-tiered incentive package for clean coal projects in Texas. Projects that capture at least 50 percent of their carbon dioxide emissions are eligible for sales tax exemptions for the equipment that captures, transports, and stores their carbon dioxide. The first three projects that can achieve a 70 percent carbon capture rate will qualify for an additional bonus, a franchise tax credit of $100 million per project.

Transportation: Reducing Vehicle Miles Traveled

As mentioned in Chapter 8, the transportation sector accounted for 30 percent of greenhouse gas emissions in 2005. Thus, one effective way of reducing greenhouse gas emissions would be to reduce vehicle miles traveled. In urban areas, public transportation is a viable option to reach this goal. Over longer distances, a modern rail system could be used to reduce truck traffic by increasing the amount of freight carried by rail. As a result of NAFTA, the quantity of goods entering the country from Mexico has increased substantially, and Texas has been building roads to transport these goods to other parts of the country.

Ever since the Second World War, urban land use planning has been dominated by suburban development. Single-family residences in suburban developments with associated shopping malls have encouraged sprawl and made residents totally dependent on their cars. In urban areas, there is now a steadily increasing interest in New Urbanism. The main principles of New Urbanism are traditional neighborhoods with higher-density, mixed-use development and a pedestrian-friendly street design. Most destinations should be within a 10-minute walk of home or work. The higher density also makes it easier to develop public transportation systems. Thus, residents are likely to reduce the number of trips that they make by automobile.

Most major cities in Texas are encouraging vertical mixed-use development, often allowing higher densities than previously permitted in exchange for a component of affordable housing. Public transportation is improving, with Dallas and Houston developing successful light rail systems. State policy still makes this option more difficult to implement than the building of roads, by requiring voter approval, even when the funding is available. In 2000, Austin voters narrowly defeated a proposal to build a light rail system. On second try, in 2004, a much more modest system utilizing an existing 32-mile line was approved. The Austin trains, which finally became operational in March 2010, are powered by diesel engines, thus limiting the reduction in greenhouse gases and local pollution that would have been obtained by electrically powered systems.

Water Management: Incorporating Climate Change

Agencies responsible for the management of natural resources—air, land, seashore, water, and wildlife—must do more to consider climate change in their management plans and operating procedures. The management of water resources and the coastal zone is used here to illustrate how adapting to global warming will change existing policies.

The Texas Water Development Board has successfully decentralized water planning. With guidance from the state, the main task of preparing the state water plan is now entrusted to water managers, experts, and citizens of 16 water basins. The basin reports are then integrated into a single report that is submitted to the Legislature (TWDB 2007a). The board should guide the regional planning groups in broadening their planning efforts. In the current planning cycle, to be completed in 2011, basin and state reports should consider climate change and climate variability alongside traditional factors, such as population growth, agriculture, urban development, biodiversity, and in-stream flow requirements. The "Science Policy" section above provides guidelines on how to structure this effort. These guidelines were used in conducting an integrated assessment of water resources in the Lower Rio Grande Valley (Houston Advanced Research Center 2000; Schmandt 2006).

Case Study: Lower Rio Grande Basin

The Lower Rio Grande Valley, a four-county area in the most southern part of Texas, will be especially vulnerable to climate change. This stems from several sources: poverty, dependence on surface water from Mexico, brackish groundwater, and lack of infrastructure. The Valley is semiarid, prone to drought, and economically dependent on its agricultural base. Although some of the acreage is restricted to rain-fed cotton and grain sorghum, 40–50 percent of the land is irrigated, enabling the production of high-value crops such as fruit and vegetables. Because of the widespread use of irrigation, agriculture accounts for 88 percent of water demand in the Lower Rio Grande Valley. Even so, the demand for water is so much greater than the supply that only in the wettest years are all of the agricultural needs met. In addition, with high birthrates and continuing immigration, municipal demand for water is expected to increase two- or three-fold by the year 2040. Because municipal use has a higher priority than agricultural use, this will be at the expense of agricultural users.

Water in the Lower Rio Grande is shared with Mexico. In 1944, Mexico agreed to supply two thirds of Lower Rio Grande water from its tributaries, in exchange for water provided by the United States to Mexico in the Colorado basin. To unravel the problem cluster linking water to development in an arid, rapidly growing region, a team of specialists from Texas and Mexico conducted an integrated assessment for the Lower Rio Grande basin. The questions to be answered were straightforward: "Will there be enough water, of acceptable quality, to support the sustainable development of the region to the year 2050? What will be the impact of climate change on water supply?" Team members had experience in hydrology, water quality, ecology, demographics, economic development, and water management. The assessment proceeded in three stages: initial scoping of issues and concerns, detailed analysis of major issues and development scenarios, integration and policy recommendations. Throughout the process, water managers and users were consulted. With few exceptions, team members used existing information and population projections, even though some data sets from Mexico and the United States were not immediately compatible. The main task was not to generate new data but to interpret and link existing information that had been produced by different disciplines and territorial jurisdictions.

Hydrological data for the drought of record in the 1950s showed a reduction of stream flow to half of average levels and rapid declines in reservoir levels during years 1 to 7 of the drought. During years 8 to 10, reservoir levels slowly recovered. Combining historical drought data with a 2°C increase in temperature and a 5 percent reduction in precipitation (a reasonable assumption about climate change impacts by 2030), the reservoirs would be exhausted by year 5, with no recovery occurring during the remainder of the 10-year drought period. This would undermine the economic and environmental sustainability of the region.

Our research identified several management options for coping with increased water shortages:

1. Irrigated agriculture currently uses 88 percent of available river water. Improvements in water distribution and use, water metering, as well as changes in crop patterns, can maintain current crop yields while reducing water use by 40 percent.
2. Urban and industrial activities use 12 percent of river water. To meet the future demands of the growing population, their share must rise to 20 percent.
3. The necessary transfer of agricultural to municipal water can be done in one of two ways: through regulatory changes, which will constrain the existing rights of water users, or through a water market, which will help suppliers and users to develop a less invasive solution.
4. The region has already suffered significant damage to aquatic and terrestrial resources. Although full restoration is unlikely, governments can still act to prevent further deterioration.
5. Desalinization of brackish groundwater or seawater is not yet cost-effective, but it is not out of reach as an emergency measure.

The assessment, completed in 2000, included a number of scenarios to evaluate the severity of possible future events. Two of these contingencies have since materialized, with a new multiyear drought and a dramatic reduction in water delivery from the main Rio Grande tributary in Mexico. The assessment showed that a combination of these two factors would lead to severe water shortages and economic losses. Indeed, farmers on both sides of the border have suffered large losses. The study showed that climate change will aggravate a repeat of the drought of record. Remedial measures, if taken in time, can reduce the risk.

Meeting Water Demand

The TWDB, the state's water planning agency, predicts that the Texas population will more than double between the years 2000 and 2060, and demand for water will increase 27 percent. Existing water supplies (the amount of water that can be produced with current permits, current contracts, and existing infrastructure during drought) are projected to decrease about 18 percent, primarily because of the accumulation of sediments in reservoirs and the depletion of aquifers. The TWDB has identified 4,500 water management strategies and projects to generate an additional 9 million acre-feet (1 acre-foot equals 1233.48 cubic meters) per year of water supplies for Texas, at a cost of about $30.7 billion (Table 10.5). According to the TWDB, if Texas does not implement the state water plan, water shortages during drought could cost businesses and workers in the state about $9.1 billion per year by 2010 and $98.4 billion per year by 2060, and leave about 85 percent of the state's projected population without enough water. None of these projections takes into account reductions in supply as a result of changes in precipitation and increases in evaporation induced by global warming.

Two general approaches may be taken to respond to the problem of decreased water availability in the state: resource expansion and resource management. Resource expansion is accomplished through structural solutions that involve either the construction of additional reservoirs to capture more water or interbasin transfers of existing supplies. These options face serious financial and environmental obstacles. An alternative approach is to manage existing supplies more effectively through conservation, water pricing, and enforcing the existing structure for allocation of supplies. Other ways of increasing surface water supplies include purchasing additional water through contracts with major water providers, obtaining additional water rights, reallocating water in existing reservoirs, and changing the operating framework for a system of reservoirs. Together these increases in surface water supplies would account for almost half of the total needed.

Each of these options could be utilized to improve on the present management system and may help the state cope with the widely varying regional effects of global warming. Although we have analyzed each of these alternatives as separate policy options, it is important to note that they are not

TABLE 10.5. Strategies for meeting Texas water demand in the year 2060

MANAGEMENT STRATEGY	WATER SUPPLY, IN MILLION ACRE-FEET	COST, IN MILLIONS OF DOLLARS
Conservation	2.04	939
New major reservoirs	1.07	4,904
Other surface water conveyance projects, water marketing, reallocation of reservoir water	3.31	13,175
Additional groundwater supplies	0.80	2,330
Reuse of wastewater	1.26	3,965
Desalination of brackish and sea water	0.31	2,590
Conjunctive use of surface and groundwater	0.23	2,800
Total	9.03	30,700

Source: Texas Water Development Board, Chapter 10 in Water for Texas, 2007.

mutually exclusive. Indeed, conservation efforts and marginal water pricing could enhance the current system of water allocation.

Structural Solutions

The TWDB, in its 2007 report to the Legislature, states that 14 major new reservoirs generating approximately 1.1 million acre-feet of supply are needed to meet future demand over the next 50 years (TWDB 2007a). Additionally, the TWDB estimates that 29 water conveyance projects will be required to carry water from new and existing reservoirs to areas of greatest demand. A number of these projects would serve the city of San Antonio. The magnitude and number of these projected structures raise serious concerns about the feasibility of this approach.

The construction of additional reservoirs faces four major obstacles: (1) the most favorable sites for reservoir construction in the state have already been developed, (2) reservoir construction often entails a 30-year lead time, (3) projects of this magnitude raise serious environmental concerns, and (4) federal funding for such projects has decreased dramatically in recent years. Similar problems face projects for interbasin transfers of water. A legal restriction on interbasin transfers represents another obstacle to such a response. The Texas Water Code mandates that interbasin transfers of surface water may be considered only for water that exceeds the 50-year water requirements of the originating basin. This restriction minimizes flexible water planning and does not recognize the disparate availability of water throughout the state. In the 2007 water plan, TWDB has included this recommendation: "The Legislature should provide statutory provisions that eliminate unreasonable restrictions on the voluntary transfer of surface water from one basin to another."

The City of El Paso, in partnership with the U.S. Army (Fort Bliss), has received federal funds to build and operate the largest inland desalination plant in the United States, capable of producing 27.5 million gallons daily of drinking water from previously unusable brackish groundwater (http://www.epwu.org/water/desal_info.html). This represents one quarter of the city's current drinking water needs. The University of Texas at El Paso opened a desalination research center with $6 million in startup funding, $2 million of which was provided by the State of Texas. The research will initially focus on reducing the amount of water lost during desalination (in some plants 50 percent, in the modern El Paso plant 15 percent) and methods for commercializing the mineral residue (*El Paso Times,* October 24, 2008). Brownsville, at the mouth of the Rio Grande, is operating a pilot plant for desalinization of sea water. A full-scale plant, estimated to cost $150 million, is in the planning stages (*USA Today,* July 1, 2007).

Water Conservation

Water conservation constitutes a valuable first response mechanism for reducing future demand for water supplies. Expected savings through conservation are in the range of 2 million acre-feet, of which two thirds would come from agriculture. Although the largest potential savings can be made in the agricultural sector, there is considerable pressure on municipalities to implement water conservation plans, particularly in areas that are using all of their existing supplies. These figures do not include water savings from legislation requiring more efficient plumbing fixtures. However valuable these efforts are, they will be insufficient in themselves to meet future water demands in the state.

Reductions in agricultural use, from 60 percent of current statewide demand to about 40 percent, will come from a number of sources. Many of the technological improvements available for delivering water to crops can save 20–30 percent of the water currently applied. They include low-energy precision application sprinklers, surge flow irrigation systems, and drip irrigation. Additional savings could be realized by better maintenance of water transmission systems, including lining canals to reduce seepage and limiting evaporation. TWDB also assumes that the amount of irrigated acreage will decrease. Over the last 20 years, this trend has already become apparent, particularly in the Lower Rio Grande Valley, where financial incentives have been used to encourage farmers to convert their agricultural water rights to municipal use.

Municipal users, including residential, commercial, and institutional customers, make up the fastest-growing use sector in Texas and have significant potential for conserving water. In 2000, daily per capita consumption in Texas cities ranged from 120 to 275 gallons. Among large Texas cities, El Paso has been most successful, lowering its water use from 149 gallons of water per person per day in 2003 to 134 gallons per person per day in 2007 (*El Paso Times*, June 14, 2008). The state recommends a conservation target of 140 gallons. Water utilities in Texas currently cannot account for a significant amount of the water they treat and distribute, as much of it is lost through leaks in distribution systems. Proper auditing techniques and modern, electrical leak detection equipment could be used to reduce transmission and distribution losses. A 2003 state law (House bill 3338) requires all retail public water utilities to submit a water audit report showing the utility system's annual water loss to the TWDB every five years. The most recent report shows statewide losses for 2005 estimated at 0.21–0.46 million acre-feet, or 5.6–12.3 percent of all water entering the reporting system (TWDB 2007b). In addition, a more recent state law greatly expands the number of retail public water utilities required to prepare water conservation plans and submit them to TWDB.

Reuse of treated effluent is predicted to increase three-fold by 2060, saving 1.26 million acre-feet. This water can be used for such purposes as industrial water supply, landscape and agricultural irrigation, direct recharge of aquifers, and other environmental uses.

Drought Management

Planning for years of less than average precipitation is an important component of water resource management in arid western states. The majority of western states have adopted drought management plans that are coordinated by the state in response to specific trigger conditions. The Palmer Drought Severity Index, which rates drought conditions on a severity scale of 1 to 5, is often used. When the index falls below 2, certain measures, often voluntary, are instituted. At this stage the main goal is to educate the public of the potential for water shortages. As drought conditions become more severe, interagency task forces are activated and specific programs implemented. As necessary, the governor will typically work with these task forces to solicit additional legislative authority and to secure financial assistance from the federal government.

In 1999, in response to recent droughts that had brought massive losses to agriculture, the Texas Legislature passed House bill 2660 on drought planning and preparation. The bill formed the Drought Preparedness Council, which is housed in the governor's Division of Emergency Management. The council advises the governor, Legislature, and state agencies on significant drought conditions. It is organized as four committees:

- Drought Planning and Coordinating Committee: responsible for statewide planning, preparation of State Drought Preparedness Plan, recommending specific revisions for a statewide response, and coordinating agencies involved.
- Drought Monitoring and Water Supply Committee: assesses and reports on meteorological and hydrological conditions and forecasts, and makes a determination concerning when to activate State Drought Response Plan.
- Drought Technical Assistance and Technology Committee: advises regional water planning groups on drought-related issues, maintains a database of water suppliers, and coordinates technical and financial assistance to drought-impacted communities.
- Drought Impact Assessment Committee: assesses and reports on potential impacts of water shortages on the public's health, safety, and welfare, on economic development, and on agricultural and natural resources.

The council issued a Drought Preparedness Plan in 2005. The responses to particular drought effects in a geographic area are initiated by agency representatives in each committee. These response actions are planned well in advance of a drought situation or, in the case of unforeseen situations, will be the result of intense analysis of available problem data by each respective agency. Additional or emergency assistance needs that cannot be met through the council's resources are passed to the governor's Division of Emergency Management for further action (Drought Preparedness Council 2005).

The council has the task of coordinating drought planning and response of 16 state agencies. It also serves as liaison with 12 federal agencies. It is difficult to judge the effectiveness of the council. An independent review of its work should be performed. To date the Drought Council is not on the schedule of upcoming agency reviews by the Texas Sunset Commission.

The State of Texas also requires public water suppliers and irrigation districts to prepare and regularly update a drought contingency plan. Oversight of the process and review of plans submitted by large water utilities and irrigation districts is a function of the Texas Commission for Environmental Quality.

Market-Based Pricing of Water

Water is typically priced at a rate that reflects the cost of operating the collection and purification systems and the distribution network. It does not include most of the capital costs associated with constructing reservoirs and conveyance infrastructure, because much of this expense has been subsidized by the federal government. If water were to be priced at its replacement cost, it would more closely approximate its market value. Taking into account the geographic region of the state, type of water use, and capital costs, an economic pricing system for water could be developed. Under this system, market forces would help to distribute the resource to those who most desire it, as indicated through their willingness to pay. Adopting marketplace pricing of water would require a fundamental cultural change and would raise controversial political questions, including the ability of poorer citizens to pay. Needless to say, the implementation of such a system would not be easy, with user groups vigorously challenging such price hikes. Agricultural users, in particular, would be hard hit, and many crops could not be grown profitably without subsidized water supplies. The economic foundation of many rural communities could be severely undermined.

Using economic factors as a means of redistributing current supplies, a market for water rights has developed in the Lower Rio Grande Valley. In 1991, active water rights for irrigation alone in the Valley exceeded 1.7 million acre-feet, well above the firm yield of the system. It is only in years of above-average precipitation that farmers are able to exercise options to the full extent of their water rights. With population growth, there has been a shift from agricultural to municipal water use, and irrigators have been able to sell or lease their rights to growing municipalities. The priority of the right changes, however, when water rights are converted from agricultural to municipal use. Because of this, an adjustment in the amount of water transferred is necessary. Presently, a municipality receives a dependable 40–50 percent of water rights purchased from the irrigation districts. This conversion formula is based on the firm annual yield of the system, so that if all the water were converted to municipal use, it would not exceed the amount of water available from the system under drought conditions.

Coastal Management: Incorporating Climate Change

Coastal vulnerabilities already exist as a result of land subsidence and coastal erosion. Addressing these problems through better land use management is a cost-effective response to rising sea levels and potential storm events. Significant portions of the Texas Gulf Coast currently suffer from land subsidence and coastal erosion, with the result that coastal land is subjected to regular flooding. The Texas coast, therefore, is a particularly useful model for studying the potential effects of rising sea levels induced by climate change. The impacts are potentially devastating for urban areas (see Chapter 7) and coastal ecosystems. Increasing demand for fresh water inland and the possibility of reductions in supplies as a result of climate change will place additional strains on the coastal environment (see Chapter 4).

Several state agencies, with overlapping and sometimes conflicting authority, share responsibility for the coastal zone. The Texas Water Commission, for example, controls waste discharges into the Gulf, while the Texas Water Development Board conducts studies on the freshwater inflow needs of the estuaries. The state owns the coastline below the high-water mark and attempts to clean beaches and remove squatters through the Texas General Land Office. The Texas Parks and Wildlife Department is responsible for the management of biological resources along the coast, the Texas Railroad Commission has exclusive authority over oil and gas wells, and the Texas Health Department is responsible for certifying shellfish for human consumption. Thus, it is not surprising that an integrated approach to management of the coast has been difficult to develop.

Coastal Engineering and Adaptation

Currently Texas has no comprehensive policies for the management of its coastline in response to climate change. There have been local efforts to mitigate some of the most dramatic impacts of shoreline erosion and storm damage, most notably the construction of a seawall on Galveston Island and the introduction of improved management practices by the Houston-Galveston Coastal Subsidence District. In general, policies that address these problems fall into two categories. The first, shoreline engineering, consists of physical modifications to the coast to hold it in its present position. The second is adaptation. Recognizing that the coast will change, the goal is to minimize damage from a rise in sea level by, for instance, limiting development in sensitive areas.

Structural Responses

Of the possible approaches to shoreline engineering, beach replenishment is the most environmentally sound. This procedure consists of pumping sand, usually dredged from off shore, onto or near the beach. There are a number of problems with this approach, not the least of which is cost. In addition, after replenishment the beach is steeper, causing waves to strike with greater force than before and accelerating the rate of erosion. Other potential problems arise because the sand used for replenishment differs from the original beach sand, and offshore dredge pits, which alter wave action, can have negative effects.

Another method to deal with erosion is to construct groins. Groins are walls built perpendicular to the coast; they are intended to capture sand carried in long shore currents. Typically, groins have worked well and sped accretion to local areas. Unfortunately, by capturing more sand locally, they speed erosion farther along the coast. Jetties also function in the same way. Like groins, jetties are constructed perpendicular to the coast, but they are several times longer. The primary purpose of jetties is to protect ship channels from silting. Although they are very successful in ensuring safe entry to and exit from harbors, jetties also capture as much as 50 percent of the sand supply that would otherwise go to Texas beaches.

The most dramatic example of shoreline engineering is the construction of seawalls. Constructed back from the shoreline, seawalls are intended primarily to protect inland property from storm damage. The obvious example is the Galveston seawall, constructed to prevent a repeat of the devastation caused by the 1900 hurricane. The cost of construction is tremendous; roughly $7 million per mile, and outside the most developed locales, like the city of Galveston, the cost of a seawall is greater than the value of the property it protects. Typically, a massive seawall like that at Galveston is part of an ongoing process that began with small bulkheads intended to mitigate the effects of occasional wave impacts. In these cases, a combination of factors results in an accelerating rate of beach erosion, and increasingly massive structures become necessary as more and more pressure comes to bear. In the end, there will be no beach left, just a huge wall overlooking the wreckage of its predecessors.

Institutional Responses

Texas has already seen some success with new types of institutional responses to coastal problems. The Harris-Galveston Coastal Subsidence District, for example, has been successful in reducing land subsidence by restricting groundwater withdrawal. Most of the areas have effectively decreased the amount of groundwater used to 10 percent of the total water demand through conversion to surface water and water conservation. Reliance on surface water for public supplies will create additional problems, however, if the availability of surface water decreases as a result of climate change.

National policies also affect coastal development. The National Flood Insurance Act of 1968 created the National Flood Insurance Program to provide low-cost flood insurance, on the condition that the community directs new development out of the hazardous area, a condition that was not effectively enforced. The Flood Disaster Protection Act of 1973 required flood insurance with any type of financial loan that was federally insured for any property in a hazardous area prone to flooding and flood-related erosion. This act also directed the Federal Emergency Management Agency to identify flood-related erosion zones. The Upton-Jones Amendment encourages the demolition or relocation of structures in hazardous areas by advancing payment. The Coastal Barriers Resource Act limits federal investment on undeveloped coastal barriers.

Texas was the thirtieth state to receive federal approval of its coastal zone management plan, 10 years after the previous applicant. It was originally submitted to the National Oceanic and Atmospheric Administration (NOAA) in October 1995, and approval by NOAA was published in the Federal Register on January 10, 1997. NOAA approval makes the state eligible for federal financial assistance, which is used to assist the state in administering the various state and local authorities included in the Texas Coastal Management Plan, as well as to fund local management efforts to increase public access, restore damaged resources, and manage coastal erosion. The General Land Office was designated as the lead agency to develop a long-term plan for the management of Texas coastal public land, in cooperation with other state agencies that have duties relating to coastal matters. Although public participants in the process identified coastal erosion and wetlands loss as areas of major concern that will be exacerbated by climate change, there is no explicit reference to global warming in the current statewide coastal management policy.

Several state coastal programs are addressing climate change issues via statewide, interagency climate change partnerships, often under their governor's climate change initiatives. The coastal programs are providing information for, or responding to, specific action items generated by these state climate commissions. In this capacity, coastal programs are playing a key role in ensuring the consideration of coastal impacts and adaptation strategies. For instance, Louisiana's Coastal Program is participating with state and nongovernmental organizations in an initiative entitled "Climate, Energy, and the Coast." The initiative is focused on the restoration of Louisiana's wetlands. Although not a response to climate change, the Texas Coastal Program is supporting development of local geohazard maps that include sea-level rise, erosion rates, wetlands, and other information, like the one developed as a planning tool for the City of Galveston by the University of Texas. A similar map is being developed, with funding under Coastal

Zone Management Act Section 309, for Mustang Island and the City of Port Aransas (Climate Change Work Group 2007).

The Coastal Zone Management Act should be recognized by Congress and the federal administration as one of the primary statutes that can foster adaptation to climate change at the state and local levels. The states' coastal programs often directly manage shoreline development and work closely with local governments on land use planning, habitat acquisition, and a variety of other activities. They also play a key role in coordinating state and local agencies and have the authority to review and condition federal permits in the coastal zone.

OUTLINE OF A COMPREHENSIVE TEXAS CLIMATE POLICY

Considering the ongoing work on energy efficiency and alternative fuels, I find that Texas is doing more than is generally acknowledged in the reviews of state carbon dioxide mitigation efforts published by the EPA, the Pew Center, and other sources. This is so because the current Texas energy strategy adds up to a hidden climate change policy. The driving forces are energy efficiency and independence, as well as the income and jobs associated with industries developing alternative energy sources. Governor Rick Perry is a strong promoter of this approach: "I want Texas to be the epicenter of energy development—wind, solar, clean coal, obviously natural gas, nuclear and biofuels" (*Austin American-Statesman*, July 18, 2008). This is very much in the tradition of policy development in Texas, to endorse economic development without paying much attention to environmental goals. Fortunately, most alternative energy programs also reduce greenhouse gas emissions, thus creating win-win situations that simultaneously advance energy security and a reduction in greenhouse gas emissions.

But more is needed. The time has come for Texas to develop a comprehensive policy that links climate change to energy independence, regional security, and the management of natural resources. Such a policy will serve the interests of Texas in several ways.

In the first place, it will make it easier to separate win-win from win-lose strategies in promoting energy efficiency and searching for new energy sources. Win-win strategies include green buildings, more stringent vehicle and appliance standards, and energy from renewable sources such as wind, sun, wood chips, switchgrass, and algae. Win-lose strategies include corn-derived biofuels, increased use of coal without investment in clean coal technologies, and more nuclear power without securing safe storage for spent nuclear fuel, to name three examples. The economic and environmental costs and benefits of these technologies must be compared to each other, so that policymakers receive reliable guidance for their decisions. In the case of biofuels from corn, for example, the amount of energy and water needed and the increase in food prices make this technology an unwise choice (National Research Council 2008). Even though evidence supporting this conclusion is strong, Texas continues to support farmers who grow corn for biofuels.

Climate and Security

Texas has begun to increase its energy independence, but other policy issues also need to be addressed. Recent research has identified climate-related security risks (Busby 2007; CNA 2007). As the author of one of these studies argues, the country "needs to 'climate proof' its domestic infrastructure including military installations, particularly along its coasts, [by] substantial investment in risk reduction: coastal defenses, building codes, emergency response plans, and evacuation strategies" (Busby 2008). I add the electric distribution network to this list. The impacts of Hurricane Ike would have been less dramatic if more electric utilities had been buried underground. Texas should not wait for federal action but needs to improve security-related infrastructure on its own. As a first step, four possible risk scenarios need to be evaluated:

- Coastal infrastructure. The large refineries, chemical plants, and transmission lines in proximity to the Gulf Coast and the Gulf Intracoastal Waterway are vulnerable to catastrophic storms and ocean surges.
- Water conflicts. International conflicts can result from decreased water supplies on the border with Mexico.
- Migration. Current climate change assessments predict severe droughts in northern Mexico. This can lead to increased trans-border migration.
- Tropical diseases. The northward march of tropical diseases such as dengue fever and water-borne diseases creates new health risks that medical doctors are not sufficiently trained to diagnose and treat.

Scenarios do not predict the future; they illustrate possible futures. As such, they are useful starting points for study and discussion. Once this is done, decision makers should review results for possible changes in policy and management (Table 10.6).

TABLE 10.6. Challenges and responses

CHALLENGE	RESPONSE
Link climate predictions to management of air shed, water basins, and ecological regions	Conduct multidisciplinary, place-based sustainability assessments; include stakeholders in assessment process
Link climate predictions to urban management	Establish urban service centers modeled after Agricultural Research and Extension Service
Regulate greenhouse gas emissions	Encourage federal regulations that do not penalize Texas for its oil and gas services to the rest of the nation
Develop alternative fuels	Support industry to become a leader in wind, solar, and biofuel energies
Develop clean energy	Support electric utilities to bring storage of CO_2 and nuclear fuel to maturity
Increase energy efficiency	Revise building codes; adopt advanced standards for appliances and vehicles
Prepare for water scarcity	Reward conservation; increase irrigation efficiency; prepare for droughts; manage water as an economic good
Prepare for sea-level rise	Develop comprehensive coastal management plan; adopt and enforce stringent zoning and insurance requirements
Increase regional security	Evaluate risks to coastal infrastructure, risks from conflicts between water users, risks from drought-induced cross-border migration, and risks from northward shift of tropical diseases
Planning and coordination	Establish State Office of Energy, Security, and Climate

Better Policy Advice

Texas needs a strong research and planning effort to prepare for climate change. Its primary task will be to provide first-class policy advice to state and regional decision makers and agencies about possible risks of climate change, win-win solutions, and nonstarters, as well as new security risks. Following the already mentioned recommendation by the National Academy of Sciences, regional climate change is best not examined as a single issue but in conjunction with other issues standing in the way of a watershed's, air shed's, or city's sustainability. Interdisciplinary research of this kind should be undertaken by Texas research universities and think tanks, with coordination and partial funding by the New Technology Research and Development Program described below. Additional funding should be solicited from federal agencies and foundations.

A Texas Office of Energy, Security, and Climate

This chapter's single most important recommendation is this: Texas should create an Office of Energy, Security, and Climate. A plan for its organization and function should be developed as a first step. The Legislature, governor, and lieutenant governor should convene a joint study committee to prepare such a plan and make recommendations on the mission, organization, and location of the office.

The office should develop a comprehensive climate policy. Much of its work should focus on opportunities that can make the state a leader in renewable energy, energy efficiency, clean coal technology, and management of drought and sea-level rise. Creation of the office will present an opportunity to assess the need for consolidation of related efforts currently located in different agencies—the Governor's Office, Comptroller's Office, Texas Commission for Environmental Quality, Railroad Commission, State Energy Conservation Office, and Drought Preparedness Council, among others.

The example of California provides useful guidance on how to proceed. Key steps in California included enabling legislation, implementation by a technically experienced agency, and coordination with other state agencies and cities. Florida, more conservative in its approach, provides another model. First, the governor convened an action-focused conference on climate change. He then appointed the 27-member Action Team on Energy and Climate Change. Based on the Action Team's recommendation, the governor signed three executive orders aimed at reducing Florida's greenhouse gases, increasing energy efficiency, and pursuing renewable energy sources. In addition, Governor Crist is partnering with Germany and the United Kingdom to discuss and promote initiatives that broaden the Kyoto Protocol and reduce the emission of greenhouse gases beyond 2012 (http://www.dep.state.fl.us/climatechange/team/default.htm; http://myfloridaclimate.com/).

The following actions should be taken by the Texas Office on Energy, Security, and Climate: preparation and updating of a greenhouse gas inventory, recommendations on statewide greenhouse gas targets and caps, preparation of a climate change action plan, consultation and possibly cooperation with regional climate change alliances, convening of a state action team on climate change, and public outreach and education. The office would also advise line agencies responsible for water, air, land, wildlife, and coastal management on how to integrate climate change into their operations. Similarly, the office should help business participate in federal low-carbon programs and transition to a low-carbon economy.

Establishment of the office will take time. In the meantime, two immediate initiatives can build on existing programs.

Texas Emerging Technology Fund

The authors of this book recommend a significant increase in state funding for development, design, and testing of clean energy, alternative energy, energy storage, and carbon dioxide capture and storage projects. The Texas Emerging Technology Fund would manage the new projects. Created by the 2007 Legislature, the fund has awarded $110 million as research, matching, and commercialization grants. So far, the main focus has been on information technology and the life sciences. A few grants deal with biofuels and energy. One of these, totaling $4 million, was awarded at the governor's urging to advance knowledge about the use of algae for biofuels (Texas Emerging Technology Fund 2008). A $2 million grant supports water desalinization research (*El Paso Times*, October 24, 2008). The fund should broaden its focus and support emerging technologies that advance energy efficiency, regional security, and greenhouse gas mitigation.

New Technology Research and Development Program

The New Technology Research and Development Program should receive additional resources to address new tasks. This program is sponsored by the Texas Commission for Environmental Quality and currently provides financial incentives to support research, development, and commercialization of technologies that reduce air pollution (http://tceq.state.tx.us/implemen tation/air/terp/program_info.html#research). The program awards $2 million each year for air quality research and $9 million for new technology development. The Texas Environmental Research Consortium, a nonprofit organization based in Houston, administers the program under contract with the Texas Commission on Environmental Quality. The program has made excellent progress in identifying the sources and pathways of air pollutants in the Houston and Dallas-Fort Worth metro areas. The research results have been synthesized for use by local, state, and national decision makers (Texas Environmental Research Consortium 2008; Olaguer et al. 2006). This coalition of universities, local stakeholders, and the state has won national recognition for its innovative work at the interface between research and action.

The New Technology Research and Development Program should expand its activities and initiate, manage, and coordinate regional assessments of climate change impacts and response strategies. As it currently does for air pollution assessments, the program would issue requests for proposals to the Texas research community, carefully select grantees, and summarize research results for policymakers. As a new step, the program could convene study groups that respond to information requests by the Legislature, governor, and state agencies. This would give Texas the institutional capacity to study and offer policy advice on issues related to energy, security, and climate. These issues will play a critical role in shaping the future of Texas. The National Research Council, a branch of the National Academy of Sciences, has long provided unbiased research assistance of this kind to the Congress and federal government. Texas policymakers should consider emulating this model and adjusting it to the regional scale.

Texas and Federal Policy

With regard to mitigation, we repeat our 1995 recommendation that Texas engage actively in the national debate about regulating carbon dioxide. Chapter 9 provides details on the impact of federal legislation on the Texas economy. Whatever national legislation is passed, Texas must watch that no disproportionate burden is imposed that would penalize the state for the large energy services it provides to the rest of the nation. To date, the Texas congressional delegation has been mostly absent from the national debate. There have been a few notable exceptions. Congressman Lloyd Doggett, in June 2008, introduced the Climate Matters Act (Climate Market Auction Trust and Trade Emissions Reduction System). Five Texas representatives—Frost, Hinojosa, Jackson-Lee, Johnson, and Reyes—have cosponsored bills with mandatory carbon limits. The governor's draft 2008 energy plan finally endorsed a stronger Texas contribution: "State policymakers should bring a Texas perspective to federal carbon policy debates. Texas needs to participate actively in the carbon discussion and educate Washington decision makers on the economic value of Texas' energy production to the nation and prevent Texans from being punished for providing the energy and petrochemical products that the rest of the nation consumes" (State Energy Plan 2008).

Identifying and financing adaptation measures will be largely the responsibility of the state and its cities and counties. Texas should aggressively pursue federal research and demonstration opportunities similar to the federal support received for the "new water in the desert" partnership between El Paso Water Utility, Fort Bliss, and the University of Texas at El Paso.

REFERENCES

Austin, 2007. Final Report to Mayor and Council of the 2015 Zero Energy Capable Homes Taskforce. Memorandum. Austin City Council, Austin, Tex.

Baker, N., 2009. Survey Shows Majority of Texans Desire More Clean Energy. Energyboom website, August 21, 2009. http://solar.energyboom.com/category/tags/cynthia-and-george-mitchell-foundation.

Barron, E., 2008. Gaining Traction beyond the Science of Climate Change. Presentation at "Beyond Science: The Economics and Politics of Responding to Climate Change," Rice University Conference, February 9, 2008, Houston. http://webcast.rice.edu/webcast.php?action=details&event=1481.

Betsill, M., 2000. *Localizing Global Climate Change: Controlling Greenhouse Gas Emissions in U.S. Cities.* J. F. Kennedy School of Government, Harvard University, Cambridge, Mass. http://www.hks.harvard.edu/gea/pubs/2000–20 .pdf, accessed June 24, 2008.

Busby, J., 2007. *Climate Change and National Security: An Agenda for Action.* Council Special Reports 32. Council on Foreign Relations, New York.

Busby, J., 2008. Insecure about Climate Change. *Washington Post,* March 22, 2008.

California Air Resources Board, 2008a. http://www.arb.ca.gov/cc/cc.htm, accessed July 2008.

California Air Resources Board, 2008b. Executive Summary, *Climate Change Draft Scoping Plan,* June. Sacramento, Calif.

City of Austin Web, 2007. Austin Climate Protection Plan. http://www.ci.austin .tx.us/council/downloads/mw_acpp_points.pdf, accessed July 9, 2008.

City of Austin Web, 2008. Austin Climate Protection Program. http://www.ci.austin .tx.us/acpp/default.htm, accessed July 9, 2008.

Climate Action Partnership, 2009. *A Blueprint for Legislative Action.* Website http: //www.us-cap.org).

Climate Change Work Group, 2007. *The Role of Coastal Zone Management Programs in Adaptation to Climate Change.* Coastal States Organization, Washington, D.C.

CNA, 2007. *National Security and the Threat of Climate Change.* Report by Military Advisory Board. CNA, Alexandria, Va. http://cna.org, accessed July 2008.

Drought Preparedness Council, 2005. State Drought Preparedness Plan. Office of the Governor, Division of Emergency Management, Austin, Tex. http://www.txwin .net/dpc/, accessed September 21, 2008.

Environmental Defense Fund, 2008. The Cap and Trade Success Story. http://www .edf.org/page.cfm?tagID=1085, accessed September 4, 2008.

Environment News Service, 2008. Texas to Spend Billions on Wind Power Transmission Lines. Press release, July 18, 2008. http://www.ens-newswire.com /ens/jul2008/2008-07-18-094.asp.

EPA, 1997. *Climate Change and Texas.* Report 230-F-97-008qq. http://yosemite.epa .gov/OAR/globalwarming.nsf/UniqueKeyLookup/SHSU5BWHN8/$File/tx _impct.pdf, accessed June 17, 2008.

Florida Energy and Climate Change Action Plan, 2008. http://www.dep.state.fl.us /climatechange/files/action_plan/intro.pdf, p. 21, accessed November 26, 2008.

Friedman, Thomas L., 2008. *Hot, Flat, and Crowded: Why We Need a Green Revolution, and How It Can Renew America.* Farrar, Straus and Giroux, New York.

Governor's Competitiveness Council, 2008. *Texas State Energy Plan.* Office of the Governor, Austin. http://governor.state.tx.us/files/gcc/2008_Texas_State_Energy _Plan.pdf.

HARC 2000. *Water and Sustainable Development in the Lower Rio Grande/Río Bravo Basin.* Final report to the U.S. Environmental Protection Agency. Houston

Advanced Research Center, The Woodlands, Tex. http://mitchell.harc.edu /Archive/RioGrandeBravo/.

Harriss, R., 2008. Testimony before the Subcommittee on Energy and Environment, U.S. House Committee on Science and Technology, Hearing "Energizing Houston: Sustainability, Technological Innovation, and Growth in the Energy Capital of the World," February 29, 2008. http://files.harc.edu/Documents /Announcements/2008/UrbanSustainability.pdf.

IPCC, 2007. Summary for Policymakers. In: *Climate Change 2007: Impacts, Adaptation, and Vulnerability.* Contribution of Working Group 2 to the Fourth Assessment Report of the Intergovernmental Panel on Climate Change. Cambridge University Press, Cambridge, U.K., and New York.

Kerry, J., 2008. The Road from Bali: The Future of American Policy on Global Climate Change. Presentation at "Beyond Science: The Economics and Politics of Responding to Climate Change," Rice University Conference, February 9, 2008, Houston, Tex. Answer to question 2. http://webcast.rice.edu/webcast .php?action=details&event=1404.

Litz, Franz T., 2008. *Toward a Constructive Dialogue on Federal and State Roles in U.S. Climate Change Policy.* White Paper, Pew Center on Global Climate Change, Arlington, Va.

Lorenz, P., D. Pinner, and T. Seitz, 2008. The Economics of Solar Power. *McKinsey Quarterly,* June 2008.

Lovins, A., 2008. Using Energy More Efficiently: An Interview with the Rocky Mountain Institute's Amory Lovins. *McKinsey Quarterly.* http://www.mckinsey- quarterly.com, accessed September 24, 2008.

Lutsey, N., and D. Sperling, 2008. America's Bottom-up Climate Change Mitigation Policy. *Energy Policy* 36(2):673–685.

Mitchell Foundation, 2008. Press release, February 6, 2008. George and Cynthia Mitchell Foundation. http://www.reuters.com, accessed July 19, 2008.

National Caucus of Environmental Legislators, 2008. Select State Actions to Address Climate Change. http://www.ncel.net/newsmanager/news_article.cgi?news_id =184, accessed April 20, 2010.

National Research Council, 1999. *Our Common Journey.* National Academy Press, Washington, D.C.

National Research Council, 2008. *Water Implications of Biofuels Production in the United States.* National Academy Press, Washington, D.C.

NOAA, 2008. *Weather and Climate Extremes in a Changing Climate.* Report by the U.S. Climate Change Science Program and the Subcommittee on Global Change Research, Washington, D.C. http://www.climatescience.gov/Library/sap/sap3-3 /final-report/default.htm.

Norwine, J., and K. John, eds. 2007. *The Changing Climate of South Texas, 1900– 2100.* CREST-RESSACA, Texas A&M University, Kingsville.

Olaguer, E., H. Jeffries, G. Yarwood, and J. Pinto, 2006. Attaining the 8-hr Ozone Standard in East Texas: A Tale of Two Cities. *EM Magazine,* Air and Waste Management Association, October, p. 26.

Pew, 2006. *Agenda for Climate Action*. Pew Center on Global Climate Change, Arlington, Va.

Pew, 2008a. *U.S. States and Regions*. Pew Center on Global Climate Change, Arlington, Va. http://www.pewclimate.org/states-regions, accessed June 23, 2008.

Pew, 2008b. State Legislation from Around the Country. Pew Center on Global Climate Change, Arlington, Va. http://www.pewclimate.org/what_s_being_done /in_the_states/state_legislation.cfm.

Potential Gas Committee, 2009. Potential Gas Committee Reports Unprecedented Increase in Magnitude of U.S. Natural Gas Resource Base. Press release. Colorado School of Mines, Golden, Colo. http://www.mines.edu.

Rabe, Barry G., 2006. *Race to the Top: The Expanding Role of U.S. State Renewable Portfolio Standards*. Pew Center for Global Climate Change, Arlington, Va.

RGGI, 2010. RGGI CO_2 Auctions Yield Millions for Investment in Clean Energy, Job Creation. Press Release, March 12. RGGI, New York. http://www.rggi.org.

Ridge Energy Storage and Grid Services, 2005. *The Economic Impact of CAES on Wind in TX, OK, and NM*. State Energy Conservation Office, Austin, Tex. http://www.seco.cpa.state.tx.us/zzz_re/re_wind_projects-compressed2005.pdf, accessed July 19, 2008.

Sanders, E., 2005. Rebuffing Bush, 132 Mayors Embrace Kyoto Rules. *New York Times*, May 14, 2005.

Schmandt, J., 2006. Bringing Sustainability Science to Water Basin Management. *Energy* 31:2014–2024.

SECO, 2008. State Energy Conservation Office, Austin, Tex. http://www.seco.cpa .state.tx.us/re_wind.htm, accessed July 11, 2008.

Sjardin, M., 2008. Cap and Trade Systems: The European Experience. Presentation at "Beyond Science: The Economics and Politics of Responding to Climate Change," Rice University Conference, February 9, 2008, Houston, Tex. Powerpoint file at http://www.rice.edu/energy/events/past/ClimateChange/cc_beyond science9feb08.html#_pres.

Sperling, D., 2008. *The California Model for Combating Climate Change*. Presentation at "Beyond Science: The Economics and Politics of Responding to Climate Change," Rice University Conference, February 9, 2008, Houston, Tex. http://webcast.rice.edu/webcast.php?action=details&event=1466.

Speth, J. G., 2004. *Red Sky at Morning: America and the Crisis of Global Environment*. Yale University Press, New Haven.

State Energy Plan, 2008. Draft report of the Texas Governor's Competitiveness Council. Office of the Governor, Austin. http://www.governor.state.tx.us/gcc, accessed July 14, 2008.

Texas Emerging Technology Fund, 2008. Texas One Program, http://www.texasone .us/site/PageServer?pagename=tetf_homepage, accessed July 14, 2008.

Texas Environmental Research Consortium, 2008. http://www.tercairquality.org/, accessed July 20, 2008.

TWDB, 2007a. *Water for Texas, 2007*. Texas Water Development Board, Austin.

TWDB, 2007b. An Analysis of Water Loss as Reported by Public Water Suppliers in Texas. Prepared by Alan Plummer Associates, Inc., for Texas Water Development Board, Austin.

United Nations Development Program, 2007. *Human Development Report 2007/2008: Fighting Climate Change*. UNDP, New York.

U.S. Conference of Mayors, 2007. *Survey on Mayoral Leadership on Climate Protection*. Washington, D.C. http://www.usmayors.org/climateprotection/climate survey07.pdf, accessed June 24, 2008.

U.S. Conference of Mayors, 2010. *Mayors Leading the Way on Climate Protection*. Washington, D.C. http://www.usmayors.org/climateprotection/revised/, accessed April 21, 2010.

U.S. Global Change Research Program, 2000. *Climate Change Impacts on the United States: The Potential Consequences of Climate Variability and Change, Overview*. Cambridge University Press, Cambridge, U.K., and New York.

Vedlitz, A., 2004. *Use of Science in Gulf of Mexico Decision Making Involving Climate Change*. U.S. Environmental Protection Agency Cooperative Agreement, EPA Project R-83023601-0. Bush School of Public Policy, Texas A&M University, College Station.

Vedlitz, A., 2008. The 21st Century Challenge: Water Policy in the Context of Climate Change. Presentation at "Forecast: Climate Change Impacts on Texas Water," conference April 28–30, 2008, Austin, Tex.

Zweibel, K., J. Mason, and V. Fthenakis, 2008. Solar Grand Plan. *Scientific American* January 2008:63–74.

Contributors

JORGE BRENNER is a postdoctoral research associate in Ecological Economics at Texas A&M University–Corpus Christi.

JUDITH CLARKSON is a self-employed research scientist in biology and public affairs.

BILL DAWSON is the editor of *Texas Climate News*. He formerly covered environmental issues for the *Houston Chronicle*.

JAMES GIBEAUT is the Endowed Associate Research Professor in Geospacial Sciences at Texas A&M University–Corpus Christi.

WENDY GORDON is the program leader for Nongame and Rare Species in the Wildlife Division of the Texas Parks and Wildlife Department.

JARED HAZLETON is Principal for Economics at Texecon.

DAVID HITCHCOCK is the Director of Sustainable Transportation Programs at Houston Advanced Research Center (HARC).

BRUCE A. McCARL is Distinguished Professor and Regents Professor of Agricultural Economics at Texas A&M University–College Station.

PAUL A. MONTAGNA is the Endowed Chair of Ecosystem Studies and Modeling at Texas A&M University–Corpus Christi.

SALLY MOREHEAD is program coordinator for Resource Management at the University of Texas Marine Science Institute–Port Aransas.

JOHN W. NIELSEN-GAMMON is the Texas State Climatologist and Professor of Meteorology at Texas A&M University–College Station.

GERALD R. NORTH is Distinguished Professor of Meteorology and Oceanography at Texas A&M University–College Station.

JANE M. PACKARD is Associate Professor of Ecology and Evolutionary Biology at Texas A&M University–College Station.

JURGEN SCHMANDT is Professor Emeritus of Public Affairs at the University of Texas at Austin and Distinguished Fellow at Houston Advanced Research Center.

GEORGE H. WARD is a research scientist in Meteorology and Hydrology at the University of Texas at Austin.

Index

285; population trend, 175, *175*;
water resource issue, 173, 181,
181–182, 286
employment, 241, 264, 277
endangered species, 134, 138, 143,
147, 165, 168
energy consumption: climate change
effects, 33–34; historical changes
in, 250n19; projected increase in,
34; reducing, 204–208, 212, 239;
shifting to renewable sources,
210–211; in Texas, 201, 202–203,
226–228, 227; urban areas, 182–
184, *183*, 192–193
energy efficiency and independence:
Austin's initiatives, 270–271; buildings, 207–208, 214, 215; incentives
for, 228, 244; need for further
initiatives, 217; as Texas focus,
7, 206–208, 276–278; vehicles,
205–206, 212–213
Energy Efficiency Building Retrofit
Program, 214
Energy Policy Act (EPACT) (1992), 244
Energy Policy and Conservation Act
(1975), 204
Energy Security and Climate
Stewardship Platform for the
Midwest, 267
energy sources. *See* alternative energy
sources; fossil fuels
Energy Star program (EPA), 184
Energy Tax Act (1978), 244
ENSO (El Niño–Southern Oscillation),
11, 30–31, 111
environmental intelligence centers, 275
Environmental Protection Agency
(EPA): air quality standards, 189;
climate change impacts, 257, *257*;
Energy Star Program, 184; fuel
economy standards, 213; greenhouse gas emissions authority, 198,
261; recognition of greenhouse gas
threat, 2, 6, 7–8, 232

EPACT (Energy Policy Act) (1992), 244
erosion: controlling, 73, 74, 289–290;
vs. sedimentation of shoreline, 102,
104, 105
estuaries: and biodiversity, 142–143,
144; freshwater inflow into, 108–
109, 118–119; Texas, 97–102, 99,
104, 107
ethanol, 206, 265, 293
European climate change initiatives,
130, 265
European Union Emission Trading
Scheme, 249n8
evaporation rates: biodiversity effects,
128, 135–136, 141, 144; coastal
areas, 96, 111; projections, 29, *30*;
and water cycle, 76; water resource
effects, 35, 50, 64, 80, 85, 90, 91,
258, 283
externality, economic, defined, 230
extinctions of species, threat of, 126,
131, 133, 146
ExxonMobil, 8

Far West Texas precipitation patterns,
40, *43*. *See also* El Paso
federal policies: California as inspira-
tion for, 264–265; coastal development, 290–291; Congressional
carbon emission control legislation,
236–237, 261; greenhouse gas
emission regulation, 2, 6, 7–8, 198,
232, 260–262; green initiatives, 2,
3, 4; incentives for alternative energy usage, 278; learning from state
and local initiatives, 268; legislative
initiatives, 235–244; need for, 216;
Texas engagement with, 297. *See
also* national level
field programs to add to data set,
19–20
financial incentives: for alternative
energy sources, 233, 234, 244,
278; for energy efficiency, 228,

water supply effects of warming, 34–35. *See also* modeling climate change

sea-level change: agricultural sensitivity to, 158; and biodiversity, 144; coastal impact, 12, 96, 102–107, *105, 107,* 117, 118; economic aspect, 229; global average rise, 28; policy considerations, 269

seasonal variations in Texas: precipitation, 59, *60,* 61, *61,* 64; temperature, 51–56, *53–55,* 64

sea surface temperatures: in Pacific and precipitation, 11, 30–31, 56, 63; and severe weather increases, 31, 111, 229

seawalls, 290

security consideration, 293

sedimentation vs. erosion of shoreline, 102, 104, 105

service sector as contributor to Texas economy, 223

severe weather impacts: agricultural sensitivity to, 158; flooding, 50, 111, 139, 185, 187; on infrastructure, 229–230; and sea-surface temperatures, 31, 111, 229; summary, 2–3, 11; and Texas climate, 48–51, *49,* 65–66; tornadoes, 48, *49,* 65; transportation infrastructure, 185–186; tropical storms, 50, 111; urban area vulnerabilities, 177; and water supply, 70, 72, 73. *See also* hurricanes

shoreline changes, 101–102, 118. *See also* sea-level change

shoreline engineering, 289–290

Smith, J., 148

solar brightness changes, 24–25

solar energy, 206, 211, 246, 250n16, 278–279

South Central Plains, 138–139

South Central Texas, precipitation in, 44, 46. *See also* San Antonio

Southeast Texas, precipitation in, 47, *47. See also* Houston

South Texas: biodiversity and climate change in, 140–141; hydroclimatological region, 77, *81, 82,* 92; population trends, 222; precipitation in, 44, 46, 47; retreating shoreline of, 101–102. *See also* coastal areas

Southwest monsoon, 34, 40

Soward, Larry, 272

species, wildlife: assisted colonization, 148; distribution of, 12, 96–97, 131–132, 142; endangered, 134, 138, 143, 147, 165, 168; extinctions threat, 126, 131, 133, 146; traits and climatic shifts, 130–131; vulnerability to climate change, 133–135

spectroscopy challenge to anthropogenic global warming, 25

Sperling, D., 262, 267

springflow, and Edwards Aquifer losses, 167–168, *168*

State Energy Conservation Office, Texas, 206–207

state level: climate change initiatives, 262–265; coastal management programs, 291–292; greenhouse gas emissions, 206–207, 208–210, *209,* 212–213, 263, *263*–266, 267–268; vehicle emission control, 205, 263. *See also* California; Texas

steam-electric power generation, 78–79, *181*

storm surge, 50, 144

Straus, Joe, 4

Structure of Scientific Revolutions, The (Kuhn), 21

subsidence on Texas coast, 96, 102, 103, 229

subtropical grassland, 140–141

surface water: future management of, 284; in hydroclimatology, 70, *81;*